W9-CKL-892

Motivation and Disposition: Pathways to Learning Mathematics

Seventy-third Yearbook

Daniel J. Brahier
Seventy-third Yearbook Editor
Bowling Green State University
Bowling Green, Ohio

William R. Speer
General Yearbook Editor
University of Nevada, Las Vegas
Las Vegas, Nevada

ISSN 0077-4103
ISBN 978-0-87353-661-5

The National Council of Teachers of Mathematics is a public voice of mathematics education, supporting teachers to ensure equitable mathematics learning of the highest quality for all students through vision, leadership, professional development, and research.

Printed in the United States of America

Contents

Preface

The 2011 National Council of Teachers of Mathematics (NCTM) Yearbook offers an opportunity and a resource to explore a wide variety of perspectives on motivation and disposition as they relate to mathematics teaching and learning. This variety includes examining such elements as the demographic composition of a school; the role of movies, television, and the Internet; and nontraditional pedagogy as means of promoting and influencing positive student and teacher dispositions. Motivation and disposition evolve (or stem) from many sources, and many factors can magnify or diminish both. The presence or lack of curiosity, ambition, parental influence, teacher encouragement, peer pressure, and future goals—these and many other elements play central roles. Against these realities, rhetoric is less effective than concrete ideas for implementation—this book is designed to offer a perspective to the reader that reinforces action.

The seventy-third yearbook of the NCTM, *Motivation and Disposition: Pathways to Learning Mathematics*, consists of twenty-one chapters divided into four major sections. The first five chapters, part I, present an overview of research and the theoretical underpinnings of motivation and disposition; the section is titled "Perspectives on Motivation and Disposition." Part II, "Cultural and Societal Issues," includes chapters 6–9 and explores cultural dimensions associated with the topic. Part III, "Motivation and Disposition in the Classroom," is a seven-chapter section that delves into practical suggestions for promoting and assessing dispositions in daily teaching. The final section, part IV, is titled "Professional Development Support" and consists of five chapters that explore the role of promoting long-term change toward valuing the role of motivation and disposition in mathematics education. Together, these twenty-one chapters present a vision and a challenge for mathematics teachers to develop positive attitudes and dispositions in their students.

Producing an NCTM yearbook would be an insurmountable task without the help and guidance of many people. One of the most challenging parts of putting together a yearbook is to select from the multitude of manuscripts submitted in response to the original call. As is the custom, the selection and organization of these manuscripts eventually falls heavily on the general editor and volume editor, but the editorial panel represents the heart and soul of the process. I am indebted to the editorial panel for the 2011 Yearbook for their deep expertise and a wide body of experiences in disposition and motivation across the spectrum of mathematics education. For serving as members of the 2011 Yearbook Editorial Panel, my special thanks go out to the following individuals for their consistent demonstration of insight, knowledge, creativity, and command of the tasks before them: Lynn Breyfogle, Melfried Olson, Marian Small, Marilyn Strutchens, and Denisse Thompson.

Several other individuals worked behind the scenes to ensure the prompt and quality production of the volume. These include Joanne Hodges, director of publications; Myrna Jacobs, publications manager; Gabe Waggoner, copy and production editor; Ken Krehbiel, associate executive director for communications; Kathe Richardson, meeting planner; and Amy Roth McDuffie and Linda Cooper Foreman, Educational Materials Committee chairpersons during the production of this yearbook. These dedicated people were there every step of the way and offered much direction and support on process and procedure. In fact, many other unnamed members of the NCTM staff worked long and hard to bring this project to fruition.

Of course, this yearbook would not have been possible without the outstanding contributions of the volume editor, Daniel Brahier of Bowling Green State University. Dan has been a beacon for us all on this project. His unwavering dedication and work ethic have helped us in ways that are most visible in the quality of the final product that will help others comprehend the roles and dynamics of motivation and disposition in teaching and learning mathematics for years to come.

William R. Speer
NCTM General Editor, 2011–2013 Yearbooks

Introduction

Motivation and Disposition: The Hidden Curriculum

Daniel J. Brahier

Consider This Situation

In addition to my university responsibilities, I also teach an eighth-grade mathematics course each morning, which I have been doing for twenty years. For their third-quarter project, students worked together in teams of four to investigate the life and contributions of famous mathematicians. The students explored individuals such as René Descartes, Euclid, Carl Friedrich Gauss, Hypatia of Alexandria, Emmy Noether, Benoît Mandelbrot, and Blaise Pascal, which gave the class a wide spectrum of mathematical topics as well as periods in history. Each group worked as a team to gather information about the mathematician, but I also assigned them three different tasks (fig. 1 below and continuing).

Mathematics History Project

Rationale

The purpose of this project is to explore the historical development of mathematical ideas by studying famous mathematicians who contributed to the knowledge that is studied in school today.

Requirements

Each group of 4 students will investigate the life and accomplishments of one famous mathematician. Exploration will be conducted on the Internet, and three different products (written paper, poster, and skit script) will be turned in. On the due date, each group will make a presentation to the class about their assigned mathematician. Each person will complete a peer evaluation form at the end of the project to rate one another's contributions.

Individual Product Items

Written Paper (75 points)

One person in the group will be assigned the role of "writer." This person will compile information about the mathematician, including his/her life story and mathematical accomplishments and write a paper about the person. The paper must be word-processed and have a cover page that includes the writer's name and the mathematician's name and picture. The paper must be 12-point, 1-inch margins, double-spaced, and be 3 pages in length (no more than 4 pages). It should include the mathematician's early life, education and mathematical work (including the effect on our world and thought), interesting details of his/her life, later life, and a bibliography of all sources used. (Use parentheses to cite sources and quotation marks if using a direct quote. Do *not* copy and paste from the Internet—plagiarism will result in a zero on the paper and demerits.) The writer will be assessed on the following:

1. Accuracy of the paper
2. Clarity of writing
3. Correctness of mathematics discussed
4. Grammar and spelling

Poster (75 points)

One person in the group will be assigned the role of "artist." This person will compile information about the life of the mathematician and create a poster about the individual. The poster must be designed on a half sheet of regular poster paper. Lettering is to be either word-processed or neatly/artistically hand drawn. The poster should "tell the story" of the mathematician at a glance to a person looking at it in the hallway. The artist will be assessed on the following:

1. Accuracy and completeness of information in the poster
2. Correctness of mathematics displayed
3. Use of color and pictures
4. Overall attractiveness of the poster (spelling, lettering, etc.)

Script (75 points)

Two people in the group will be assigned the role of "playwright." These people will compile information about the life of the mathematician and his/her accomplishments and write a script for a 3- to 5-minute skit that will be performed by all of the team members on the due date. The skit should be a "scene" from the life of the mathematician that illustrates a mathematical scene the person was in, famous problem s/he solved, or some other appropriate situation. The script must include a narrator part where one person narrates at the beginning to set up the scene for the other "actors," as well as include a narration at the end that tells the class more about the life and accomplishments of the mathematician. The playwrights will share score and will be assessed on the following:

1. Accuracy of the contents of the script, including length (at least 3 minutes)
2. The mathematics of the scene
3. Creativity of spoken lines and suggested props
4. Flow and grammar/spelling of the script

Group Presentation (25 points)

On the due date, all team members will participate in a short presentation about their assigned mathematician. In the first 3–5 minutes, the group will put on the skit written by the playwrights and practiced by the team. The skit must be 3–5 minutes in length (practice to make sure it's the right length). This time includes any parts read by the narrator before and after the skit. After the skit, the artist will briefly describe what is included in her/his poster. The complete presentation for each group should be about 8–10 minutes, which includes a couple of minutes for answering questions. The group will be assessed on the following:

1. Quality of the skit (well practiced, people know their parts, etc.)
2. Knowledge about the mathematician as demonstrated through the skit and any questions answered afterward

Fig. 1. Mathematics history project description

The assignment required one person (the writer) to write a three-to-four-page paper about the mathematician and his or her contributions to the field. Another student was assigned the role of artist and designed a poster—suitable for hanging in the school hallway (and eventually in my mathematics methods classroom on campus)—that told the story of the mathematician. The other two students worked together to develop a script for a three- to five-minute skit that the whole team would perform for the class, featuring some situation from the individual's life. The playwrights were to be creative so that the class would be engaged in the presentation, rather than having someone merely read a paper to them.

This time was my second conducting the mathematics history project with my class, and it was a major hit both times. After completion and assessment of the project, students had to write a journal entry evaluating the usefulness of the project, including recommendations for how to improve it in the future. The following are comments from my students' journals:

- This project was cool because I take for granted the stuff we are learning and realized someone invented and/or discovered this. This was my favorite project this year, plus I probably learned the most from this project.
- I learned *so* much doing this. This project really helped me to get a view on some of the things that we take for granted in math (for example the coordinate grid). What was really cool about Hypatia was to see how math was thought about even back in the 300s, almost 2,000 years in the past.

- My favorite mathematician was Emmy Noether, because I found her story fascinating. She really had a lot going against her, but never did it break her. She is inspirational.
- I liked learning about a more modern day female mathematician. It was cool to see her contributions . . . I learned about how difficult it was for women to get into the math world. It was inspiring to read about how she pursued her goals until she got where she wanted to be.
- I was amazed to find out how even very old mathematicians found lasting discoveries that benefit math today. Benoit Mandelbrot discovered fractal geometry which not only makes beautiful pictures but allows doctors to find the surface areas of internal organs.

I was pleased about the success of this project and shared the requirements and results with my mathematics methods class on campus one morning. After class, one of my preservice teachers approached me and said, "I really like the idea of that project and am amazed at what your students came up with. But I have one question: which objectives in the state standards did this project actually address? I didn't see any in the project description and wondered how you justify doing this kind of project if it doesn't directly hit the standards." Her question and comment took me aback. Here I was, the professor who requires my teacher candidates to become intimately familiar with our state's standards and attach every learning activity in their lesson plans to a measureable outcome, yet I had just completed a project with my own junior high school class that, by design, did not directly address *anything* in the standards.

The key idea here may be "directly" addressing the standards. I believe that learning mathematics transcends acquiring skills and even developing conceptual understanding—that, beyond these goals, my role as a mathematics teacher is to spark the interests of my students and excite them about mathematics and its usefulness in our world. Also, students need to develop some historical sense of how mathematics has evolved and continues to change over time. No specific objective in my course of study may say this, but without student interest in and recognition of applying mathematics in our society, teaching them the necessary skills and competencies will be an uphill, if not impossible, task. Let's explore in a little more detail what motivation and disposition involve.

What Does It Mean to Be Motivated?

Martin Ford, now a senior associate dean at George Mason University, wrote a landmark book titled *Motivating Humans: Goals, Emotions, and Personal Agency Beliefs* (1992), in which he presented a working definition of *motivation*. He argued that motivation has three dimensions: goals, emotions, and self-efficacy.

"Goals" refers to whether the individual engages in an activity because some external reward or punishment is associated with it or simply because the person believes that the activity is inherently useful and has a desire to learn from the experience. So, we might argue on one hand that a student's goal for learning about perimeter of polygons could have an external motivation because the student recognizes that a test grade depends on it. On the other hand, that same student may have an internal goal for taking guitar lessons because learning to play the guitar is not a required or graded task. The motivation for learning to play guitar comes from the student's own drive to learn more about playing music. One goal of education in general might be to move students from setting external goals to more internal goals, helping them recognize that the mathematics they are learning is useful and worth learning, regardless of whether doing so involves grades or other devices.

Motivation in the context of emotions refers to how interest and curiosity can drive an individual to want to learn about a particular topic. As educators, we recognize that a student who is interested in a particular topic is more likely to engage in the work. This is why teachers are encouraged to write a "motivational" springboard or engagement activity for a lesson plan. If, in the first five minutes of a lesson, we can convince the students that the topic for the day connects with their interests, the class is more likely to become motivated to learn. Notice how the preceding student quotations from journal entries refer to the mathematics history project as "cool" and "fascinating"; the students found the team's work to be interesting and, therefore, motivating.

The third dimension of motivation, personal agency beliefs, refers to the idea that an individual must feel that he or she is capable of succeeding at a task to engage in it. How many times have we seen students simply give up before they even attempt an assignment because they are convinced that they can't do it? By building students' sense of self-efficacy (self-confidence), we help them believe that they can succeed and, therefore, make them more willing to engage in a mathematical task.

Motivation has many other definitions and facets, as several papers in this volume will explore, but Ford's definition may give us a starting point. When we hear someone say, "That student just isn't motivated to learn in math class," we must push ourselves to think about what that means. The framework here is to analyze that student's motivation level by thinking about the three dimensions of goals, emotions, and personal agency beliefs. Is it that the student does not believe that the topic is useful in and of itself and will engage in it only if doing so involves reward and punishment? Is the student lacking any interest in or curiosity about the topic? Or is the student simply feeling powerless because the concept appears too difficult to understand? If we can identify the root of the lack of motivation, we can address it directly through planning and teaching.

What Is a Disposition in Mathematics?

A body of literature regarding the attitudes and dispositions of mathematics students (and their teachers) also exists. Figure 2 shows various components involved and some of their relationships.

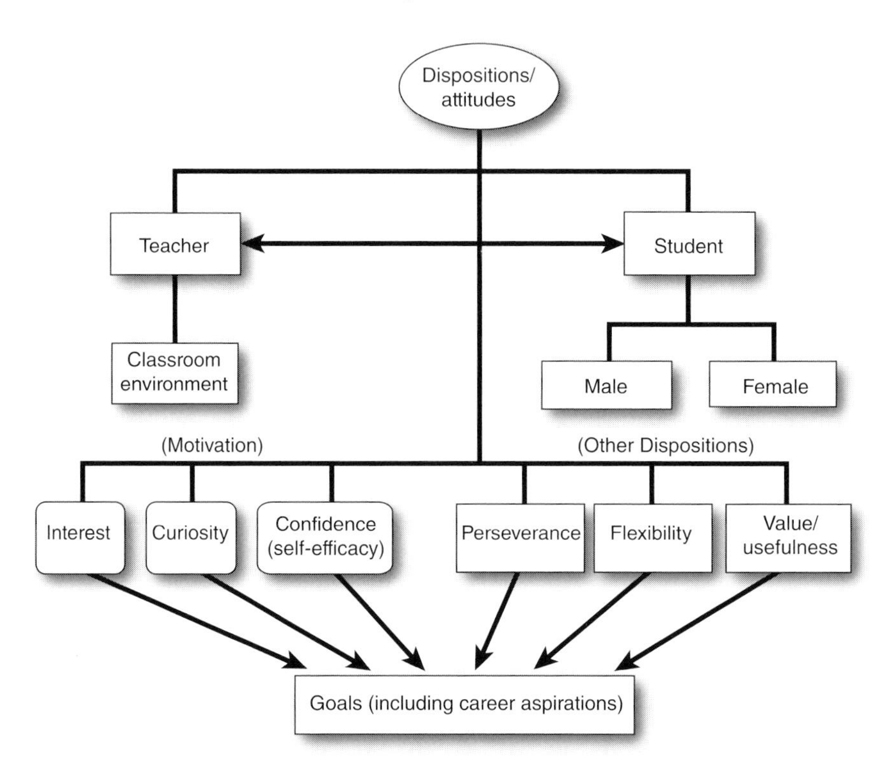

Fig. 2. Model of dispositions and attitudes in mathematics

When the National Council of Teachers of Mathematics (NCTM) released *Curriculum and Evaluation Standards for School Mathematics* (1989), it proposed an evaluation standard called "mathematical disposition." According to the document, disposition includes attitudes but is much broader than that. It includes interest in mathematics, curiosity, perseverance, confidence in using mathematics, flexibility in exploring mathematics and attempting different strategies to solve problems, valuing the application of mathematics, and appreciation of the role of mathematics in our culture and our world. NCTM stated that "disposition refers not simply to attitudes but to a tendency to think and to act in positive ways," adding that "this kind of information is best collected through informal observation" (NCTM 1989, p. 233). Further, according to *Principles and Standards for School Mathematics* (NCTM 2000), "when challenged with *appropriately chosen tasks* [emphasis added], students become confident in their ability to tackle difficult problems, eager to figure things out on their own, flexible in exploring mathematical ideas and trying alternative solution paths, and willing to persevere" (p. 21). Finally, the National Research Council cited productive disposition as one of the five strands of mathematical proficiency in *Adding It Up: Helping Children Learn Mathematics* (Kilpatrick, Swafford, and Findell 2001). The document defined disposition as "the inclination to see mathematics as sensible, useful, and worthwhile, coupled with a belief in one's diligence and efficacy" (p. 116).

The model in figure 2 places mathematical dispositions and attitudes at the top and indicates that teachers and students each possess a variety of dispositions. The dispositions associated with motivation (interest, curiosity, and self-efficacy) are at the lower left, whereas other, more general, dispositions (perseverance, flexibility, and value of the usefulness of mathematics) are at the lower right.

In studying motivation, researchers have often explored the dispositions of interest, curiosity, perseverance, and confidence (e.g., Hidi 1990; Schunk 1991; Bandura 1993). Together with flexibility and valuing the usefulness of mathematics, the dispositions that NCTM outlined directly affect goals that the student sets. These goals, then, help to shape eventual career interests and aspirations of the students. The graphic also shows that teacher and student dispositions have the potential to interact with one another. Also, teacher dispositions toward mathematics can influence classroom learning environments, and student dispositions can differ by gender. For example, females tend to believe that males are not better at mathematics but that males *think* that they are better and tend to have higher levels of self-confidence (Cramer 1989; Yong 1992). This yearbook discusses and analyzes similar research results.

Implications for Teaching

When we view teaching mathematics as a whole, it becomes clear that it is a much broader endeavor than simply helping students to acquire skills and problem-solving strategies. We are also attempting to develop motivation and positive dispositions toward studying mathematics that will have long-term effects on everything from students' confidence to do mathematics to their career choices. Each lesson in a mathematics classroom should take into account students' motivation level and dispositions and have as a goal the development of these affective characteristics. So, although the mathematics history project did not have specific standards-based objectives, it did address the interest and curiosity of my students and, I hope, gave them a historical perspective that helped them to appreciate the development and usefulness of mathematics in our society.

This yearbook is intended to give a variety of perspectives on developing dispositions in school mathematics. From research-based theory to practical ideas on how to promote productive dispositions in the classroom, the book can be a helpful tool for any mathematics teacher who strives to address the hidden curriculum—motivating students and making them eager to continue their study of mathematics.

REFERENCES

Bandura, Albert. "Perceived Self-Efficacy in Cognitive Development and Functioning." *Educational Psychologist* 28 (March 1993): 117–48.

Cramer, Roxanne Herrick. "Attitudes of Gifted Boys and Girls Towards Math: A Qualitative Study. *Roeper Review* 11 (March 1989): 128–31.

Ford, Martin E. *Motivating Humans: Goals, Emotions, and Personal Agency Beliefs*. Newbury Park, Calif.: Sage, 1992.

Hidi, Suzanne. "Interest and Its Contribution as a Mental Resource for Learning." *Review of Educational Research* 60 (Winter 1990): 549–71.

Kilpatrick, Jeremy, Jane Swafford, and Bradford Findell, eds. *Adding It Up: Helping Children Learn Mathematics*. Washington, D.C.: National Academies Press, 2001.

National Council of Teachers of Mathematics (NCTM). *Curriculum and Evaluation Standards for School Mathematics*. Reston, Va.: NCTM, 1989.

———. *Principles and Standards for School Mathematics.* Reston, Va.: NCTM, 2000.

Schunk, Dale H. "Self-Efficacy and Academic Motivation." *Educational Psychologist* 6 (June 1991): 207–31.

Yong, Fung Lan. "Mathematics and Science Attitudes of African-American Middle Grade Students Identified as Gifted: Gender and Grade Differences." *Roeper Review* 14 (March 1992): 136–40.

Part I

Perspectives on Motivation and Disposition

PART I of this yearbook, titled "Perspectives on Motivation and Disposition," consists of five papers that lay out philosophical and theoretical perspectives related to this yearbook's focus. The perspectives in these papers not only define motivation and disposition on the basis of research but also suggest factors inherent in these important constructs that can guide teachers as they make instructional decisions to engage all students in learning mathematics successfully. For readers not familiar with the research literature on motivation or disposition, these papers include an approachable, concise synthesis of important research in these domains as they relate specifically to mathematics. The foundational background one obtains through a careful reading of these papers supports the ideas in the rest of the yearbook.

"Motivation and Self-Efficacy in Mathematics Education" situates self-efficacy, namely, an individual's perceptions of his or her ability to learn, in a broader social cognitive theory that considers the reciprocal nature of personal, behavioral, and environmental factors. The paper also summarizes research about self-regulated strategy use, social and environmental models, and self-evaluations of progress as they relate to enhancing motivation and self-efficacy. The paper concludes with examples of instructional actions that teachers can use to improve their students' self-efficacy and motivation.

"A Model for Mathematics Instruction to Enhance Student Motivation and Engagement" outlines a multidimensional framework for strategies to enhance students' engagement and motivation with mathematics. In particular, the authors describe a wheel with eleven factors that are either adaptive to enhance motivation or maladaptive and impeding to reduce motivation and engagement. Although the suggested strategies focus on middle grades, they apply across the K–12 spectrum and beyond.

"Identity Development: Critical Component for Learning in Mathematics" focuses on the importance of adolescent identity development in fostering the attitudes that facilitate academic achievement. The paper includes specifics about four classes of teacher actions that are essential to developing student identity. A classroom vignette helps illustrate these teacher actions in the context of a typical classroom.

"Recommendations from Self-Determination Theory for Enhancing Motivation for Mathematics" connects recommendations from the National Council of Teachers of Mathematics *Principles and Standards for School Mathematics* with three crucial components of self-determination theory: autonomy, competence, and relatedness. The authors share research that indicates how fulfilling these three psychological needs can lead to intrinsic motivation toward learning mathematics. With minor adjustments, teachers can modify many common instructional practices to address self-determination theory's three components.

The final paper in this section, "Student Dispositions with Respect to Mathematics: What Current Literature Says," summarizes research about mathematical dispositions as well as dispositions toward mathematics. In particular, the authors consider cognitive, affective, and conative (volitional) mental functions and how each relates to dispositions. Their summary affords a common language that mathematics educators can use in discussing the role that dispositions play in mathematics achievement.

While reading the papers in this section, consider the following questions:

- As teachers, why should we be concerned about our students' motivation or disposition toward learning mathematics?
- How does instruction aligned with the *Principles and Standards* relate to student motivation and disposition toward mathematics? How might such aligned instruction enhance student motivation?
- On the basis of the perspectives in these papers, what current actions in your classroom contribute to enhancing students' motivation and disposition? What actions, if any, in your classroom might impede students' motivation and disposition?
- What insights do these papers offer about how the instruction or environment in your school building or department could enhance or impede students' motivation to learn mathematics?
- What might teachers gain from explicit discussions with students about factors related to motivation and disposition? What do these papers suggest about the interaction of teachers' expectations of students' learning and students' motivation to learn with dispositions toward mathematics?
- What do these papers suggest about your own motivation and disposition toward learning mathematics?

Denisse R. Thompson

Chapter 1

Motivation and Self-Efficacy in Mathematics Education

Dale H. Schunk
Kerri Richardson

CONSIDER THE following teacher actions:

- Mrs. Anderson, a first-grade teacher, opens each mathematics lesson with a variety of visual activities to help her students warm up for the day's explorations. One such activity involves ten frames (Wheatley and Reynolds 2010). She displays the ten frame (fig. 1.1) for a few seconds and asks students to describe what they saw. One student says, "I saw 6 dots and 4 empty spaces." Another says, "I saw 4 empty spaces, so that means there must be 6 dots." Still another student notes, "I saw 2 dots and 4 more, which is 6." Mrs. Anderson carefully listens to all the configurations that her students saw and then displays the image for everyone to view. Other hands go up, and students describe even more ways to view the ten frame. She then transitions to the lesson of the day by asking more questions about what students know about the particular topic.

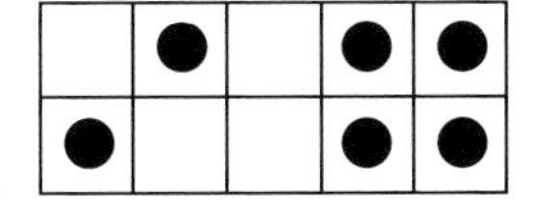

Fig. 1.1. Mrs. Anderson's ten frame

- Mrs. Swanson, a third-grade teacher, distributes worksheets for students to practice their multiplication facts at the beginning of each mathematics lesson. She sets a timer, and students have an allotted time to fill in all the necessary blanks. The timer sounds, and with no discussion Mrs. Swanson collects the worksheets and begins the mathematics lesson for the day.

Which activity would motivate students, and which would probably hinder the lesson? Which do you use in your own teaching? The activity in Mrs. Anderson's room is a classroom favorite because it encourages students to rely on their own ways of thinking in constructing ten as an abstract unit. Mrs. Anderson's transition to the mathematics lesson of the day is typically smooth, and students are eager to explore mathematics. Mrs. Swanson's activity, however, is boring and repetitious and encourages students to go through the motions of learning their multiplication facts without developing their understanding of multiplication.

These scenarios illustrate a debate in mathematics education between teaching concepts and teaching facts and procedures. The National Council of Teachers of Mathematics (NCTM) presented a unifying view. The NCTM Learning Principle states, "Students must learn mathematics with understanding, actively building new knowledge from experience and prior knowledge" (NCTM 2000, p. 20). NCTM recommends that mathematics education address facts, procedures, and conceptual understanding. Mathematics education should develop learners who set goals, assess their learning progress, and exert greater control over their learning. These latter objectives describe learners who are motivated to learn mathematics and engaged in their learning and teachers who are motivated to help students learn.

We address these objectives in this paper on motivation, the process of instigating and sustaining goal-directed activities (Pintrich 2003). We focus on one essential motivational process: self-efficacy, or one's perceived capabilities to learn or perform actions (Bandura 1997). Self-efficacy is an important variable in Bandura's (1986, 1997) social cognitive theory. An increasing body of research supports the point that self-efficacy affects individuals' mathematical motivation and achievement. Although this paper concentrates on student motivation, self-efficacy also is relevant to teacher motivation, which we address briefly.

We initially discuss fundamental points of social cognitive theory to include how self-efficacy fits in the theory. We explain influences on and consequences of self-efficacy, along with important social cognitive motivational processes. We also summarize some research that highlights the influence of instructional, social–environmental, and personal factors on self-efficacy during mathematics learning and the predictive utility of self-efficacy. The paper concludes with suggestions to enhance self-efficacy and motivation in mathematics classrooms.

Social Cognitive Theory

This section explores facets of social cognitive theory, including reciprocal interactions, self-efficacy, and motivational processes.

Reciprocal Interactions

Social cognitive theory views human functioning within a model of reciprocal influence among personal, behavioral, and social–environmental factors (fig. 1.2; Bandura 1986, 1997). Illustrating the person → behavior link, much research shows that self-efficacy (a personal factor) influences achievement behaviors such as task choice, persistence, effort, and skill acquisition (Schunk and Pajares 2009). As students work on tasks, they note progress toward their learning goals. Such progress indicators convey to them that they can perform well and enhance their self-efficacy for continued learning (behavior → person).

Person ↔ Behavior

Person ↔ Social–Environment

Social–Environment ↔ Behavior

Fig. 1.2. Reciprocal interactions in social cognitive theory

With respect to the person → social–environment link, research on students with learning disabilities shows that individuals in students' environments (e.g., teachers, other students) may react to them on the basis of attributes often typical of students with learning disabilities (e.g., low self-efficacy) rather than on the basis of their actual capabilities (Licht and Kistner 1986). In turn, feedback can affect self-efficacy (social–environment → person), such as when a teacher tells a student, "I know you can do this," which can increase the student's self-efficacy.

The behavior–environment link is apparent in many instructional sequences. When a teacher directs students' attention to a display (e.g., "Look at this."), students may attend to it with little conscious deliberation (social–environment → behavior). But students' behaviors also can affect the instructional environment (behavior → social–environment). If the teacher asks questions and students give answers that indicate a lack of understanding, the teacher may reteach some points rather than continue with the lesson.

Self-Efficacy

Self-efficacy is a belief about what one can learn or do; it is not the same as knowing what to do (Schunk and Pajares 2009). In assessing their self-efficacy, people reflect on their skills and capabilities to translate those skills into actions. In social cognitive theory, self-efficacy is a pivotal variable contributing to people's sense of agency that they can control important events in their lives (Bandura 1986, 1997).

Self-efficacy has diverse effects in achievement situations (fig. 1.3; Schunk and Pajares 2009). It can affect choice of activities. Persons with low self-efficacy may avoid tasks, whereas those who feel self-efficacious are more eager to participate (as evident with Mrs. Anderson's students). Self-efficacy also can affect effort, persistence, and learning. Students who feel efficacious about learning expend greater effort to succeed, persist longer on difficult tasks, and learn more.

Effects of Self-Efficacy
- Choice
- Effort
- Persistence
- Learning

Sources of Self-Efficacy Information
- Actual performances
- Vicarious experiences
- Social persuasion
- Physiological indicators

Fig. 1.3. Self-efficacy's effects and sources in achievement situations

People acquire information about their self-efficacy from their performances, vicarious (nonperformance) experiences (e.g., observation of models), forms of social persuasion, and physiological indicators (e.g., heart rate, sweating; fig. 1.3; Schunk and Pajares 2009). Actual performances offer the best information because they show people what they can do. But students receive much information from knowledge of how others perform. Although later personal failures can negate a vicarious increase in self-efficacy, observing success in others who are similar (e.g., peers) can raise self-efficacy in observers for performing well, because they are apt to believe that since others can learn, they can as well. Students have far greater opportunities to acquire vicarious self-efficacy in Mrs. Anderson's class than in Mrs. Swanson's.

Students often receive persuasive information that they can succeed, such as when teachers tell them, "Try harder; I know you can do it." Such feedback can raise self-efficacy, but later performance success validates the persuasive message. Students also receive self-efficacy information from physiological indicators. When learners notice that they are experiencing less stress while taking a test, they may feel more efficacious for doing well on the test.

Motivational Processes

Social cognitive theory postulates that goals and self-evaluations of progress, values, outcome expectations, and self-efficacy are important motivational processes (fig. 1.4).

- Goals and self-evaluations of progress
- Values
- Outcome expectations
- Self-efficacy

Fig. 1.4. Motivational processes in social cognitive theory

Goals and self-evaluations of progress

Goals enhance motivation and achievement through their effects on self-evaluations of goal progress and self-efficacy. Initially individuals must commit to attaining their goals because, without commitment, goals do not influence performance (Locke and Latham 2002). As learners engage in a task, they compare their current performance with the goal. Positive self-evaluations of progress raise self-efficacy and sustain motivation. A perceived discrepancy between present performance and the goal may create dissatisfaction, which can enhance effort.

But goals by themselves do not automatically motivate and raise performance. The goal properties of specificity, proximity, and difficulty are important (Locke and Latham 2002). Goals that incorporate specific performance standards are more likely to activate self-evaluations than are general goals (e.g., "do your best"). Specific goals better describe the amount of effort that success requires and promote self-efficacy because evaluating progress toward a specific goal is straightforward. Similarly, proximal, short-term goals motivate better than distant goals because assessing progress against the former is easier. Moderately difficult goals promote motivation and achievement better than goals that are overly easy or very difficult (Locke and Latham 2002) because people may procrastinate with easy goals and not attempt difficult goals.

Asking students to memorize and recall multiplication facts (Mrs. Swanson's class) can be an easy goal for some students and a difficult one for others. It is an easy goal for students who are good memorizers, who might enjoy the challenge of finishing multiplication worksheets quickly. For such learners, however, this is a low-level goal of efficiency rather than a challenging goal of developing multiplicative thinking. Conversely, memorizing facts is difficult for learners who struggle with memorization. The anxiety of completing multiplication worksheets might discourage them from engaging meaningfully in other aspects of mathematics. Many adults can relate negative experiences about having memorized times tables as children.

Multiplication in reform-based classrooms can reflect moderately difficult goals and motivating tasks when teachers help students make connections first from repeated addition and then by exploring array and area models. Within these explorations the skill of memorizing facts may come into play, but doing so is only part of the larger picture. In a conceptual approach to multiplication, students have a variety of experiences with multiplicative thinking, thus staying within the boundaries of moderately difficult goals. Although involving different mathematical skills, Mrs. Anderson's class reflects this type of conceptual approach.

Another motivational distinction is between learning goals and performance goals (Pintrich 2003). A learning goal refers to what knowledge, behavior, skill, or strategy students are to acquire, such as learning how to represent word problems in algebraic notation. A performance goal denotes what task students are to complete, such as completing a set of word problems. These goals can have different effects on motivation. A learning goal focuses students' attention on processes and strategies that help them learn and improve their skills. The task focus motivates behavior and directs attention to task aspects that are essential for learning. Students pursuing a learning goal are apt to feel self-efficacious for attaining it. In contrast, a performance goal focuses attention on completing tasks. Such goals do not highlight the importance of the processes and strategies underlying task completion or raise self-efficacy for acquiring skills. Performance goals also can lead students to socially compare their work to that of others, which can lower self-efficacy if they believe that they are performing more poorly. Mrs. Anderson's class reflects learning goals, whereas Mrs. Swanson's seems more performance goal oriented.

Values

Value refers to the perceived importance or usefulness of learning. A premise of social cognitive theory is that individuals' actions reflect their values (Bandura 1986). Learners strive to bring about what they desire and to avoid outcomes that are inconsistent with their values. They are motivated to learn and perform when they deem that learning or performance important. Students may value mathematics for various reasons, such as wanting to become mathematics teachers or believing that it has many uses in everyday life.

Achievement motivation research has shown that expectancies for success (analogous to self-efficacy) and values affect motivation (Wigfield and Eccles

2002). Expectancies predict actual achievement, and values are better predictors of choices, such as intentions to take future courses and actual enrollment in those courses (Wigfield and Eccles 1992, 2002).

Outcome expectations

Outcome expectations are beliefs about the anticipated consequences of actions (Bandura 1997)—for example, "If I study hard, I should do well on the test." People hold beliefs about the likely consequences of actions on the basis of their experiences and observations of others (models). People act in ways that they believe will be successful and attend to models who teach them skills that they value. Outcome expectations motivate behavior over long periods when individuals believe that their actions eventually will produce desired outcomes.

Self-efficacy

Students enter learning settings with various goals, which may focus on learning or performance. Students differ in how they perceive the value of learning, and they hold expectations about the consequences of various actions. Students also hold an initial sense of self-efficacy for learning, which largely reflects their prior experiences (actual and observed) with the same or similar content.

As students engage in learning activities, instructional, social–environmental, and personal variables affect their self-efficacy and motivation (fig. 1.5). Instructional variables include teachers, types of instruction and feedback, materials, and equipment (e.g., graphing calculators, technology). When students' experiences with instructional variables are positive, they are likely to experience higher self-efficacy and motivation for learning. Thus, when teachers present a self-regulation strategy for accomplishing a problem-solving task, students may experience heightened self-efficacy for learning.

Social–environmental variables include factors such as models, room conditions, distractions, and ongoing events. Observing models can inform and motivate; students who observe a similar peer succeed may believe that they can as well. Students also may socially compare their work with that of others, and the results of their assessments can raise or lower their motivation. An environment conducive to learning can raise self-efficacy and motivation, whereas distractions and competing activities may lead learners to doubt how well they can learn in that environment.

Personal variables include those student variables associated with learning, such as their perceptions of their learning, goals, self-evaluations of progress, and decisions about strategies to use. Self-efficacy and motivation are enhanced when students believe that they are learning, making goal progress, and using effective learning strategies.

Instructional
- Teachers
- Instruction
- Feedback
- Materials
- Equipment

Social–environmental
- Models
- Room conditions
- Distractions
- Ongoing events

Personal
- Perceived learning
- Goals
- Self-evaluations of progress
- Learning strategies

Fig. 1.5. Variables affecting self-efficacy and motivation during learning

Although these variables focus on student motivation, they also are appropriate for teachers. Teachers enter learning activities with goals, values, outcome expectations, and a sense of self-efficacy for promoting student learning. What happens during learning activities affects their self-efficacy and motivation as they assess their goal progress and determine whether they need to adjust their teaching strategies.

Our discussion is limited and does not include all variables that can affect motivation. For example, self-determination theorists postulate that people strive to be autonomous and engage in activities because they want to (Reeve, Deci, and Ryan 2004). Underlying human behaviors are psychological needs for competence, autonomy, and relatedness (i.e., to belong to a group). Such needs undoubtedly are present during learning and can affect motivation. Mrs. Anderson's class better fulfills these needs than does Mrs. Swanson's. Similarly, students have affective (emotional) experiences during learning; these can influence motivation, although values and self-efficacy reflect affective reactions to some degree. Attribution theory postulates that students attempt to determine why certain outcomes resulted (Weiner 1985, 2005)—for example, why they did well or poorly on a mathematics text. These attributions, or perceived causes of outcomes, can affect their self-efficacy for continued learning. Thus, students who attribute success on a test to ability and effort (e.g., "I'm good at math and I studied hard for the test") should feel efficacious about continuing to do well. Conversely, students who attribute poor performance to low ability (e.g., "I'm not good at math") are apt to experience lower self-efficacy and motivation for learning.

Research Evidence

This section summarizes some mathematics learning research from the instructional (self-regulated strategy use), social–environmental (models), and personal (goals and self-evaluations of progress) areas. We also present some research findings on the predictive utility of self-efficacy.

Self-Regulated Strategy Use

Self-regulation refers to the process whereby students activate and sustain cognitions, behaviors, and affects that are systematically oriented toward attaining goals (Zimmerman 2001). Much research attests to the benefits on motivation and achievement from students' using self-regulatory processes during learning (Zimmerman 2001).

Teaching students to self-regulate their performances by using learning strategies improves their mathematics learning and achievement (Fuchs et al. 2003). Research also has shown that applying self-regulatory strategies during mathematics learning can enhance motivation.

Lan (1998) taught graduate students self-monitoring skills during an introductory statistics course and gave them expected learning goals (e.g., identifying discrete and continuous variables). Most students had not had a mathematics or statistics class for several years. Students recorded the frequency and duration of their self-monitoring activities related to each goal. These activities included listening to lectures, reading texts, completing assignments, participating in discussions, and receiving tutoring. Students also judged their self-efficacy for attaining each goal. Self-monitoring students displayed higher achievement, self-efficacy, and motivation than did students assigned to instructor-monitoring and no-monitoring conditions.

Schunk and Gunn (1986) gave low-achieving elementary-level children modeled instruction and practice on long division that included self-regulatory use of task solution strategies. Self-efficacy and use of effective strategies most strongly affected achievement. The largest attributional influence on achievement was due to effort attributions for success (i.e., they performed well and attributed this result largely to working hard).

Using high school students, Pajares and Kranzler (1995) found that mathematical ability and self-efficacy strongly affected achievement and that ability affected self-efficacy. With college students, Pajares and Miller (1994) showed that mathematics self-efficacy was a better predictor of achievement than were mathematics self-concept, perceived usefulness of mathematics, prior experience with mathematics, and gender.

Models

Models inform and motivate. By observing models, students learn skills and strategies. They also experience a vicarious sense of self-efficacy for learning, which successful problem solving later validates (Schunk 1995).

Schunk (1981) assigned children who had experienced difficulties learning mathematical skills to either a modeling or a didactic instructional condition. Children in the modeling group observed an adult model verbalize and implement long division problem-solving steps while applying them to problems. Children in the didactic group reviewed step-by-step problem solutions. In both conditions, conducted over several days, children practiced solving problems. Although both groups increased self-efficacy, persistence, and achievement, students under the modeling condition displayed higher achievement.

Observing models whom observers perceived as similar to themselves has yielded benefits for mathematical self-efficacy and achievement. Some research has compared the effects of mastery models with those due to coping models. Mastery models demonstrate competent performance throughout the modeled sequence, whereas coping models initially experience difficulties (e.g., make errors, verbalize negative statements) but then verbalize coping statements and eventually verbalize and perform as well as mastery models. Children who have experienced mathematical difficulties may view themselves as more similar in competence to coping models.

Using addition and subtraction of fractions with like and unlike denominators—a task at which participating children previously had not been successful—Schunk, Hanson, and Cox (1987) found that observing peer coping models raised children's self-efficacy and achievement more than did observing peer mastery models. In contrast, Schunk and Hanson (1985) found that observing peer mastery models or peer coping models raised children's self-efficacy and achievement in subtraction with regrouping better than did observing adult models or no models, but the peer mastery and coping conditions did not differ. Perhaps these children, who previously had succeeded in subtraction without regrouping, drew on those experiences and believed that since their peers could learn, they could as well. Coping models may be more beneficial when students have little task familiarity or have had previous learning difficulties.

Benefits also have come from self-modeling, which involves video recording performances for children to then view. Self-modeled performances can raise self-efficacy because they convey to observers that they are becoming more competent. Schunk and Hanson (1989) used this approach. Self-modeling children displayed higher self-efficacy, motivation, and self-regulated strategy use than both children whose performances had been recorded but who did not view their videos and those whose performances were not recorded.

Goals and Self-Evaluations of Progress

Several studies involving mathematics learning found positive effects of goal properties for motivation and achievement. Giving children a proximal goal during subtraction learning raised self-efficacy, motivation, interest, and achievement more than did giving them a distant goal or a general goal (Bandura and Schunk 1981). During a long-division instruction program, giving children specific goals enhanced self-efficacy more than providing no specific goals (Schunk 1983).

Allowing students to set goals also can raise motivation, presumably because students commit to attaining the goals that they set for themselves (Locke and Latham 2002). In one study (Schunk 1985), students with learning disabilities received subtraction instruction and practice. Children set their own goals, had goals assigned, or did not set or receive goals. Self-set goals led to the highest self-efficacy and subtraction achievement.

Relative to performance goals, learning goals can exert stronger effects on motivation and achievement. In two studies (Schunk 1996), children received instruction and practice on fractions, along with either a learning goal (learning how to solve problems) or a performance goal (solving problems). In the first study, half the students in each goal condition evaluated their problem-solving capabilities. The learning goal with or without self-evaluation and the performance goal with self-evaluation raised self-efficacy, motivation, and achievement better than did the performance goal without self-evaluation. In the second study, all students evaluated their progress in learning. The learning goal led to higher motivation and achievement than did the performance goal. Schunk and Ertmer (1999) replicated these findings with college students: students who received a learning goal and an opportunity to evaluate their learning progress experienced enhanced self-efficacy and learning of computer skills.

Predictive Utility of Self-Efficacy

Researchers have explored how well self-efficacy predicts motivation and learning. Research using various mathematical content areas and participants of different ages shows that self-efficacy predicts achievement, persistence in problem solving, and interest in mathematics (Isiksal and Askar 2005; Marsh, Byrne, and Shavelson 1988; Meece, Wigfield, and Eccles 1990; Pajares and Schunk 2001; Pietsch, Walker, and Chapman 2003; Relich, Debus, and Walker 1986).

College students' self-efficacy for problem solving and for being successful in mathematics-intensive courses related positively to aspirations for occupations that were more mathematics and science intensive, such as engineering (Lent, Brown, and Gore 1997). High school students' mathematics self-efficacy related positively to interest and grades in mathematics courses (Lopez et al. 1997). Mathematics self-efficacy relates to interest in mathematics and to intentions to enroll in mathematics courses (Lent et al. 2001). Their results showed that self-efficacy strongly predicted interest.

Teacher Self-Efficacy

Teacher (or instructional) self-efficacy refers to perceived capabilities to help students learn (Tschannen-Moran, Woolfolk Hoy, and Hoy 1998). Teacher self-efficacy, like student self-efficacy, can influence choice of activities, effort, persistence, interest, and achievement (Ashton and Webb 1986; Bandura 1997; Woolfolk Hoy, Hoy, and Davis 2009). Teachers with low self-efficacy may avoid planning activities that they believe exceed their capabilities, not persist with students having difficulty learning, expend little effort to help students, and not reteach in ways that students might comprehend. Teachers with higher self-efficacy are more likely to design challenging activities, help students succeed, and persist with students having problems. Such teachers encourage student learning through better planning and interactions with students. We should expect that Mrs. Anderson's self-efficacy is higher than Mrs. Swanson's. Teachers with higher self-efficacy were likely to have a positive classroom environment, support students' ideas, and address students' needs (Ashton and Webb 1986). Teacher self-efficacy was a significant predictor of student achievement. Woolfolk and Hoy (1990) obtained comparable results with preservice teachers.

Beginning teachers' self-efficacy may actually decline over their first year, but this decline is reversible (Woolfolk Hoy, Hoy, and Davis 2009). Building teachers' self-efficacy requires opportunities for them to collaborate and their supervisors' attention to instruction. Also crucial is collective teacher self-efficacy, or teachers' beliefs that their collective capabilities can promote student outcomes (Goddard, Hoy, and Woolfolk Hoy 2000). Collective teacher self-efficacy grows stronger when teachers successfully work together to implement changes (e.g., in the mathematics curriculum), learn from one another and from other successful schools, receive encouragement for change from administrators and professional development sources, and work together to cope with difficulties and alleviate stress (Goddard, Hoy, and Woolfolk Hoy 2004). Teacher self-efficacy, individual and collective, may help contribute to retaining mathematics teachers in the profession (Woolfolk Hoy, Hoy, and Davis 2009).

Implications for Instruction

Social cognitive theory and research on self-efficacy suggest ways that teachers can improve students' mathematical self-efficacy and motivation. We briefly discuss three (fig. 1.6).

- Use high-interest activities
- Have students set goals and evaluate progress
- Teach self-regulatory skills

Fig. 1.6. Ways to improve self-efficacy and motivation

Use High-Interest Activities

Motivation and self-efficacy benefit from high-interest instructional activities. Students with higher self-efficacy show more interest in those activities than students who doubt their capabilities (Bandura and Schunk 1981; Lent et al. 2001). But interest also can build self-efficacy. Students who find an activity interesting are likely to be motivated to learn and engage in it better, which raises self-efficacy. Mrs. Anderson's high-interest activity should result in better student engagement and higher self-efficacy and motivation than Mrs. Swanson's worksheets.

Not only activities but also grouping can develop interest. Many students enjoy opportunities to work in small groups more than whole-class instruction or individual seatwork. Combining cooperative learning with metacognitive instruction (e.g., self-checking, progress monitoring) led to the highest mathematical achievement among eighth graders compared with cooperative learning alone and with individualized learning plus metacognitive instruction (Kramarski and Mevarech 2003). Small-group learning enhanced college students' mathematics learning, attitudes, and persistence in courses (Springer, Stanne, and Donovan 1999). Using collaborative groups in mathematics classes creates peer models and can raise students' individual and collective self-efficacy, motivation, and achievement.

Have Students Set Goals and Evaluate Progress

Teachers can help students set mathematics learning goals and teach them ways to evaluate their progress. Goals that motivate the best are specific and relatively close at hand. Students also can learn to divide longer-term objectives into shorter-term goals. By keeping records (written or computerized) of their goal progress, students have tangible indicators that they are learning and improving their mathematical skills.

For example, a teacher might break a unit on quadratic equations into factoring, using the quadratic formula, and completing the square. In turn, the teacher might divide each unit into subgoals with measurable indicators of completion. As students work on the subgoals, they can record when they satisfactorily complete them. Such clear progress indicators help to strengthen motivation and self-efficacy.

Teach Self-Regulatory Skills

Teachers can incorporate self-regulatory skills into their mathematics lessons. For example, teachers might assist students to learn and use learning strategies, both general and specific to the content being studied. In mathematics, useful general strategies include setting goals, self-monitoring progress, and drawing a sketch showing information given and required. Specific strategies pertain to the particular content, such as a strategy for graphing parabolas. Students who understand how to use a strategy are apt to feel a sense of self-efficacy for being successful. Students also can learn to monitor their performances, such as by keeping records of the amount of time that they spend on homework and of their scores on homework and quizzes (Zimmerman, Bonner, and Kovach 1996). Students' perceptions that they are spending time studying mathematics and that their performances are improving foster effort and ability attributions and build self-efficacy.

Another useful self-regulatory strategy is help seeking. Middle school students who perceived themselves as more competent in mathematics were less likely to avoid seeking help when they needed it than were students less competent in mathematics (Ryan and Pintrich 1997). Students need encouragement to seek teacher or peer assistance, such as when they do not understand a mathematical concept after instruction and practice. Knowing that they can seek help without penalty should make students feel more efficacious about learning.

Conclusion

Improving students' and teachers' mathematical motivation and achievement requires that they develop skills and self-efficacy for applying those skills and continuing to improve. Although motivation is complex, research supports the idea that self-efficacy can affect choices, effort, persistence, and achievement. Although most self-efficacy research has addressed computational and procedural skills, researchers increasingly are investigating motivation and self-efficacy with higher-order mathematical concepts and thinking. We believe that raising students' and teachers' self-efficacy can lead to better teaching and learning, more capable mathematics students, and a stronger mathematics teacher workforce.

REFERENCES

Ashton, Patricia T., and Rodman B. Webb. *Making a Difference: Teachers' Sense of Efficacy and Student Achievement.* New York: Longman, 1986.

Bandura, Albert. *Social Foundations of Thought and Action: A Social Cognitive Theory.* Englewood Cliffs, N.J.: Prentice Hall, 1986.

———. *Self-Efficacy: The Exercise of Control.* New York: Freeman, 1997.

Bandura, Albert, and Dale H. Schunk. "Cultivating Competence, Self-Efficacy, and Intrinsic Interest through Proximal Self-Motivation." *Journal of Personality and Social Psychology* 41, no. 3 (1981): 586–98.

Fuchs, Lynn S., Douglas Fuchs, Karin Prentice, Mindy Burch, Carol L. Hamlett, Rhoda Owen, and Katie Schroeter. "Enhancing Third-Grade Students' Mathematical Problem Solving with Self-Regulated Learning Strategies." *Journal of Educational Psychology* 95 (June 2003): 306–15.

Goddard, Roger D., Wayne K. Hoy, and Anita Woolfolk Hoy. "Collective Teacher Efficacy: Its Meaning, Measure, and Impact On Student Achievement." *American Educational Research Journal* 37 (Summer 2000): 479–507.

———. "Collective Efficacy Beliefs: Theoretical Developments, Empirical Evidence, and Future Directions." *Educational Researcher* 33 (April 2004): 3–13.

Isiksal, Mine, and Petek Askar. "The Effect of Spreadsheet and Dynamic Geometry Software on the Achievement and Self-Efficacy of 7th-Grade Students." *Educational Research* 47 (November 2005): 333–50.

Kramarski, Bracha, and Zemira R. Mevarech. "Enhancing Mathematical Reasoning in the Classroom: The Effects of Cooperative Learning and Metacognitive Training." *American Educational Research Journal* 40 (Spring 2003): 281–310.

Lan, William Y. "Teaching Self-Monitoring Skills in Statistics." In *Self-Regulated Learning: From Teaching to Self-Reflective Practice*, edited by Dale H. Schunk and Barry J. Zimmerman, pp. 86–105. New York: Guilford Press, 1998.

Lent, Robert W., Steven D. Brown, Bradley Brenner, Sapna Chopra, Timothy Davis, Regine Talleyrand, and V. Suthakaran. "The Role of Contextual Supports and Barriers in the Choice of Math/Science Educational Options: A Test of Social Cognitive Hypotheses." *Journal of Counseling Psychology* 48 (October 2001): 474–83.

Lent, Robert W., Steven D. Brown, and Paul A. Gore Jr. (1997). "Discriminant and Predictive Validity of Academic Self-Concept, Academic Self-Efficacy, and Mathematics-Specific Self-Efficacy." *Journal of Counseling Psychology* 44 (July 1997): 307–15.

Licht, B. G., and J. A. Kistner. "Motivational Problems of Learning-Disabled Children: Individual Differences and Their Implications for Treatment." In *Psychological and Educational Perspectives on Learning Disabilities*, edited by Joseph K. Torgesen and Bernice W. L. Wong, pp. 225–55. Orlando: Academic Press, 1986.

Locke, Edwin A., and Gary P. Latham. "Building a Practically Useful Theory of Goal Setting and Task Motivation: A 35-Year Odyssey." *American Psychologist* 57 (September 2002): 705–17.

Lopez, Frederick G., Robert W. Lent, Steven D. Brown, and Paul A. Gore Jr. "Role of Social-Cognitive Expectations in High School Students' Mathematics-Related Interest and Performance." *Journal of Counseling Psychology* 44 (January 1997): 44–52.

Marsh, Herbert W., Barbara M. Byrne, and Richard J. Shavelson. "A Multifaceted Academic Self-Concept: Its Hierarchical Structure and Its Relation to Academic Achievement." *Journal of Educational Psychology* 80 (September 1988): 366–80.

Meece, Judith L., Allan Wigfield, and Jacquelynne S. Eccles. "Predictors of Math Anxiety and Its Influence on Young Adolescents' Course Enrollment Intentions and Performance in Mathematics." *Journal of Educational Psychology* 82 (March 1990): 60–70.

National Council of Teachers of Mathematics (NCTM). *Principles and Standards for School Mathematics.* Reston, Va.: NCTM, 2000.

Pajares, Frank, and John Kranzler. "Self-Efficacy Beliefs and General Mental Ability in Mathematical Problem-Solving." *Contemporary Educational Psychology* 20 (October 1995): 426–43.

Pajares, Frank, and M. David Miller. "Role of Self-Efficacy and Self-Concept Beliefs in Mathematical Problem-Solving: A Path Analysis." *Journal of Educational Psychology* 86 (June 1994): 193–203.

Pajares, Frank, and Dale H. Schunk. "Self-Beliefs and School Success: Self-Efficacy, Self-Concept, and School Achievement." In *Self Perception*, edited by Richard J. Riding and Stephen G. Rayner, pp. 239–65. Westport, Conn.: Ablex, 2001.

Pietsch, James, Richard Walker, and Elaine Chapman. "The Relationship among Self-Concept, Self-Efficacy, and Performance in Mathematics during Secondary School." *Journal of Educational Psychology* 95 (September 2003): 589–603.

Pintrich, Paul R. "A Motivational Science Perspective on the Role of Student Motivation in Learning and Teaching Contexts." *Journal of Educational Psychology* 95 (December 2003): 667–86.

Reeve, Johnmarshall, Edward L. Deci, and Richard M. Ryan. "Self-Determination Theory: A Dialectical Framework for Understanding Sociocultural Influences on Student Motivation." In *Big Theories Revisited*, edited by Dennis M. McInerney and Shawn Van Etten, pp. 31–60. Greenwich, Conn.: Information Age Publishing, 2004.

Relich, Joseph D., Ray L. Debus, and Richard Walker. "The Mediating Role of Attribution and Self-Efficacy Variables for Treatment Effects on Achievement Outcomes." *Contemporary Educational Psychology* 11 (July 1986): 195–216.

Ryan, Allison M., and Paul R. Pintrich. "'Should I Ask for Help?': The Role of Motivation and Attitudes in Adolescents' Help Seeking in Math Class." *Journal of Educational Psychology* 89 (June 1997): 329–41.

Schunk, Dale H. "Modeling and Attributional Effects on Children's Achievement: A Self-Efficacy Analysis." *Journal of Educational Psychology* 73 (February 1981): 93–105.

———. "Developing Children's Self-Efficacy and Skills: The Roles of Social Comparative Information and Goal Setting." *Contemporary Educational Psychology* 8 (January 1983): 76–86.

———. "Participation in Goal Setting: Effects on Self-Efficacy and Skills of Learning-Disabled Children." *Journal of Special Education* 19 (Fall 1985): 307–17.

———. "Self-Efficacy and Education and Instruction." In *Self-Efficacy, Adaptation, and Adjustment: Theory, Research, and Applications*, edited by James E. Maddux, pp. 281–303. New York: Plenum, 1995.

———. "Goal and Self-Evaluative Influences during Children's Cognitive Skill Learning." *American Educational Research Journal* 33 (Summer 1996): 359–82.

Schunk, Dale H., and Peggy A. Ertmer. "Self-Regulatory Processes during Computer Skill Acquisition: Goal and Self-Evaluative Influences." *Journal of Educational Psychology* 91 (June 1999): 251–60.

Schunk, Dale H., and Trisha P. Gunn. "Self-Efficacy and Skill Development: Influence of Task Strategies and Attributions." *Journal of Educational Research* 79 (March–April 1986): 238–44.

Schunk, Dale H., and Antoinette R. Hanson. "Peer Models: Influence on Children's Self-Efficacy and Achievement." *Journal of Educational Psychology* 77 (June 1985): 313–22.

———. "Self-Modeling and Children's Cognitive Skill Learning." *Journal of Educational Psychology* 81 (June 1989): 155–63.

Schunk, Dale H., Antoinette R. Hanson, and Paula D. Cox. "Peer-Model Attributes and Children's Achievement Behaviors." *Journal of Educational Psychology* 79 (March 1987): 54–61.

Schunk, Dale H., and Frank Pajares. "Self-Efficacy Theory." In *Handbook of Motivation at School*, edited by Kathryn R. Wentzel and Allan Wigfield, pp. 35–53. New York: Routledge, 2009.

Springer, Leonard, Mary Elizabeth Stanne, and Samuel S. Donovan. "Effects of Small-Group Learning on Undergraduates in Science, Mathematics, Engineering, and Technology: A Meta-Analysis." *Review of Educational Research* 69 (Spring 1999): 21–51.

Tschannen-Moran, Megan, Anita Woolfolk Hoy, and Wayne K. Hoy. "Teacher Efficacy: Its Meaning and Measure." *Review of Educational Research* 68 (Summer 1998): 202–48.

Weiner, Bernard. "An Attributional Theory of Achievement Motivation and Emotion." *Psychological Review* 92 (October 1985): 548–73.

———. "Motivation from an Attributional Perspective and the Social Psychology of Perceived Competence." In *Handbook of Competence and Motivation*, edited by Andrew J. Elliot and Carol S. Dweck, pp. 73–84. New York: Guilford Press, 2005.

Wheatley, Grayson H., and Anne M. Reynolds. *Coming to Know Number: A Mathematics Activity Resource for Elementary School Teachers*. 2nd ed. Bethany Beach, Del.: Mathematics Learning, 2010.

Wigfield, Allan, and Jacquelynne S. Eccles. "The Development of Achievement Task Values: A Theoretical Analysis." *Developmental Review* 12 (September 1992): 265–310.

———. "The Development of Competence Beliefs, Expectancies for Success, and Achievement Values from Childhood through Adolescence." In *Development of Achievement Motivation*, edited by Allan Wigfield and Jacquelynne S. Eccles, pp. 91–120. San Diego: Academic Press, 2002.

Woolfolk, Anita E., and Wayne K. Hoy. "Prospective Teachers' Sense of Efficacy and Beliefs about Control." *Journal of Educational Psychology* 82 (March 1990): 81–91.

Woolfolk Hoy, Anita, Wayne K. Hoy, and Heather A. Davis. "Teachers' Self-Efficacy Beliefs." In *Handbook of Motivation at School*, edited by Kathryn R. Wentzel and Allan Wigfield, pp. 627–53. New York: Routledge, 2009.

Zimmerman, Barry J. "Theories of Self-Regulated Learning and Academic Achievement: An Overview and Analysis." In *Self-Regulated Learning and Academic Achievement: Theoretical Perspectives*. 2nd ed., edited by Barry J. Zimmerman and Dale H. Schunk, pp. 1–38. Mahwah, N.J.: Erlbaum, 2001.

Zimmerman, Barry J., Sebastian Bonner, and Robert Kovach. *Developing Self-Regulated Learners: Beyond Achievement to Self-Efficacy.* Washington, D.C.: American Psychological Association, 1996.

Chapter 2

A Model for Mathematics Instruction to Enhance Student Motivation and Engagement

Janette Bobis
Judy Anderson
Andrew Martin
Jenni Way

I like the way my Year 8 mathematics teacher doesn't just mark my work right or wrong—she actually explains what I need to do to improve. She gives me confidence so I know I can do this!

—Thirteen-year-old female student

An underlying message of the National Council of Teachers of Mathematics (NCTM) *Principles and Standards for School Mathematics* (2000) is the necessity to build students' confidence and motivation in mathematics. Aligned with other international mathematics reform literature (e.g., Australian Education Council 1990), *Principles and Standards* advocates teaching practices presumed to enhance motivation and engagement because they not only are considered desirable outcomes by themselves but also are a means to enhance student achievement. A substantial body of research now confirms that many teaching strategies designed to increase student motivation at school, particularly in mathematics, also improve learning outcomes (Stipek et al. 1998). These findings are especially relevant for teachers of students in the middle grades. The middle grades coincide with a notable decline in students' engagement levels with school and with mathematics in particular (Martin 2007, 2008). It is also a time when progression of learning is prone to lose its momentum compared with that in elementary school (Hill, Holmes-Smith, and Rowe 1993). Therefore, giving teachers guiding frameworks—to help them (and their students) develop a deeper understanding of motivation and the relationships between certain teaching practices and student achievement, motivation, and engagement in mathematics—is essential. This chapter focuses on a multidimensional framework to articulate instructional strategies that will enhance middle-grades students' engagement and motivation in mathematics.

What Is Motivation?

Holmes (1990) described motivation as the "fuel" for mathematics learning, viewing it as an "essential component of mathematics instruction" (p. 101) along with effective teachers and quality teaching. Motivated students have the energy and drive to learn and the thoughts and behaviors that reflect this energy and drive (Martin 2007, 2010). Accordingly, if educators wish to address all aspects of student motivation, they must focus on both the mind—student beliefs and expectations—and their actions, such as students' abilities to plan, manage, and persist at tasks such as homework and challenging problems.

Requests from teachers or parents to study harder or to engage more with class activities can be meaningless if students do not understand what motivation and engagement really mean in the school context and lack explicit strategies for achieving them. Furthermore, a common understanding among parents, educators (making such requests of their children), and the students themselves of what these concepts mean, and of specific strategies and advice to improve student motivation, would be helpful.

Some questions arise: How can knowledge of a theoretical framework for motivation and engagement help address our concerns about student motivation in mathematics? What teaching strategies might motivate middle-grades students to engage more in mathematics? Before looking at specific teaching strategies, exploring important aspects of motivation and engagement as part of a unified framework is useful.

The Motivation and Engagement Wheel

Motivation's sheer diversity and fragmentation is one of its limitations (Murphy and Alexander 2000; Pintrich 2003). Educational research does not always yield useful applications, and a need exists to combine advances in scientific understanding with applied utility. Accordingly, previous work has recommended giving more attention to "use-inspired basic research" (Stokes 1997; Pintrich 2003). With this in mind, Martin (2007, 2008) developed the Motivation and Engagement Wheel to capture an integrative framework that represents seminal motivation and engagement theory. The Motivation and Engagement Wheel emerged through an attempt to bridge the gap between diverse educational theorizing and practitioners' (e.g., teachers, counselors, psychologists) needs to operate within a parsimonious educational framework that they can also clearly communicate to students. Two conceptual bases underlie the wheel: (1) the general, comprising adaptive cognition, adaptive behavior, impeding/maladaptive cognition, and maladaptive behavior, and (2) the specific, comprising eleven factors subsumed under the general. Figure 2.1 shows the wheel. Above the horizontal axis are the adaptive cognitive and behavioral factors. Below the axis are the impeding/maladaptive cognitive and behavioral factors. Thus, clockwise from left to right are adaptive cognition, adaptive behavior, impeding/maladaptive cognition, and maladaptive behavior.

Although the wheel's general dimensions (adaptive cognition, adaptive behavior, impeding/maladaptive cognition, maladaptive behavior) have been important for research and theory, its specific dimensions—that is, the eleven factors of multidimensional motivation and engagement—are particularly relevant to practitioners. Pintrich (2003) proposed seven questions to guide the development of motivational science. These questions underscored the importance of articulating a model of motivation from psychoeducational theory related to self-efficacy, attributions, valuing, control, self-determination, goal orientation, need achievement, self-regulation, and self-worth. According to Martin (2007), these concepts offered a useful means for identifying the eleven specific factors for the wheel.

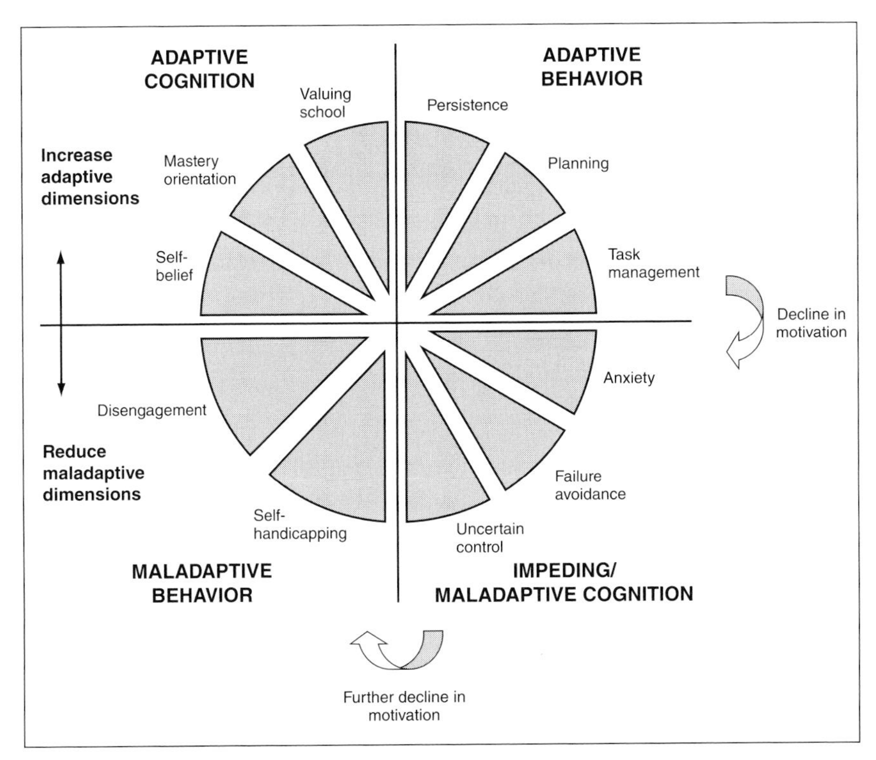

Fig. 2.1. Motivation and Engagement Wheel (Martin 2010, p. 9; reproduced with permission)

As Martin (2007) discusses, (*a*) the wheel's self-efficacy dimension reflects self-efficacy theory (e.g., Bandura 1997); (*b*) the valuing school dimension reflects valuing (e.g., Wigfield and Eccles 2000); (*c*) the mastery orientation dimension reflects self-determination (in the context of intrinsic motivation) and motivation orientation (e.g., Dweck 1986); (*d*) the planning, study management, and persistence dimensions reflect self-regulation (e.g., Zimmerman 2002); (*e*) the uncertain-control dimension reflects attributions and control (e.g., Weiner 1984); and (*f*) the failure avoidance, anxiety, self-handicapping, and disengagement dimensions reflect the need achievement and self-worth theories (e.g., Covington 1992).

Conceptualizing the wheel into the adaptive, impeding, and maladaptive dimensions allows educators to identify a set of common factors that underpin particular students' motivation and engagement, as well as which aspects teachers need to target for intervention. The practitioner and student can easily separate the helpful (adaptive) motivation from the unhelpful (impeding and maladaptive). Thus, this model, although representing a complex aggregation of theory, is an easy way for students to understand their motivation and a convenient way for practitioners to explain it to them. When students understand motivation and the dimensions that make it up, intervention is more meaningful to them and therefore is likely to be more successful.

The adaptive dimensions of the framework reflect enhanced motivation. Hence, we need students to develop self-efficacy in their abilities to understand the mathematics taught and believe that more effort will positively affect their learning. Students who break down difficult or seemingly insurmountable tasks into more achievable chunks realize that they can control and manage their learning more successfully. Thus, they are more likely to persist when they face challenging problems.

Factors in the maladaptive and impeding dimensions reflect reduced motivation and engagement. For example, feelings of anxiety can emerge when students do not understand content, are unsure how to start a task, or consider a task too daunting to even begin. Alternatively, they may use self-handicapping strategies such as doing things other than homework first or leaving the challenging academic tasks until they are too tired to do them effectively. Continued use of such behaviors can cause students to disengage from mathematics. Students will express feelings of giving up and helplessness, possibly withdrawing from the subject and even school in general unless something breaks the downward spiral.

The wheel is only the first step in addressing student motivation. The second step is to assess students on each dimension of the wheel, which is why Martin (2008) developed the Motivation and Engagement Scale (MES). The MES comprises forty-four items (four items for each specific factor in the wheel) and has proven to be a reliable and valid means of assessing students' motivation and engagement—including the role of motivation and engagement in important educational processes and outcomes such as academic (including mathematics) achievement, educational attainment, and pedagogy (Martin 2007, 2010). Together, the wheel and the MES offer a basis for educational practice and intervention aimed at more structured approaches to enhancing student motivation. Following are some guidelines and practical teaching strategies that derive from this framework.

Promoting Motivation in Middle-Grades Mathematics Classrooms

A National Research Council–Mathematics Learning Study Committee report proposed that mathematical proficiency consists of five integrated competencies: conceptual understanding, procedural fluency, strategic competency, adaptive reasoning, and productive disposition (Kilpatrick, Swafford, and Findell 2001). The report defines productive disposition as "habitual inclination to see mathematics as sensible, useful, and worthwhile, coupled with a belief in diligence and one's own efficacy" (p. 116). The notion of productive disposition aligns well with the adaptive cognition and adaptive behavior quadrants of the Motivation and Engagement Wheel, identifying such factors as valuing school, self-efficacy, persistence, and task management as crucial contributors to positive inclinations toward engaging with learning. How can teachers promote these constructive factors in the middle-grades mathematics classroom? How can we guard against maladaptive behaviors and impeding cognition, such as uncertain control and anxiety? The following section answers these questions by presenting four examples of approaches to teaching and highlighting ways to select the mathematical content, nurture positive classroom relationships, and encourage productive learning behaviors.

Variety in Teaching Approaches

The argument for using variety in teaching approaches to promote motivation and engagement begins with accepting students' individuality. Teachers need to allow for a range of personalities, learning preferences, modes of expression, and work rates. The argument continues with the notion that consistently using the same teaching approach advantages some students and disadvantages others according to the compatibility of the approach with their learning preferences. Injecting variety into teaching methods, resources used in mathematics lessons, and assessment strategies can accommodate a range of student preferences.

One can achieve variety in teaching methods by using various combinations of teacher-centered and student-centered approaches. Teacher-centered strategies include worked examples, explication, demonstration, and structured questioning. Student-centered strategies include collaborative group work, practical tasks, problem solving, open tasks, investigation, games, and student presentations.

Variety in resources involves selecting different modes of task and information delivery, as well as different materials and tools to model and explore mathematical concepts and processes. Such resources include multimedia, computer software, concrete manipulatives, models, simulations and experiment apparatus, calculators and data loggers, newspapers, excursions, and outdoor activities.

Variety in assessment strategies means using a range of formal and informal opportunities for students to demonstrate understanding before, during, and after learning events. These strategies include verbal explanation, visual presentations, concrete models, written records, tests and quizzes, work sample portfolios, self-assessment, and mastery checklists. Because assessment gives not only information for the teacher but also feedback to the student, it can affect both intrinsic and extrinsic motivation (Wiliam 2007).

Table 2.1 presents examples of three teaching methods, three resources, and three assessment strategies for teaching about multiplying decimals by powers and multiples of 10. Although we have aligned the teaching methods, resources, and assessment strategies to create three distinct teaching approaches, teachers could reconfigure the components in many productive combinations. Adjusting teaching and assessment approaches to add variety and accommodate various learning styles in these ways can develop students' self-beliefs in their capacity for success, particularly with individualized tasks and formative feedback, giving a clear direction for further learning.

Real and Relevant Tasks

Opportunities exist across the mathematics curriculum for the teacher to choose a context for tasks. The data/statistics strand of the curriculum offers an excellent example for content selection that is both relevant to the students and based in a real-life context. Content selection begins with getting to know the students. The teacher must find out about students' current interests and concerns inside and outside school, such as music, sports, movies, and local and world events. Every classroom has students who have developed considerable expertise in topics that capture their interest. Using existing interests and context knowledge not only can support engaging those particular students in the mathematical aspects of the statistical task but also creates the opportunity to hold other students' attention as they discover nonmathematical information while developing mathematical ideas and skills. Therefore, interaction between the students, within thoughtfully composed cooperative groups, becomes an important implementation factor. For example, in their Australia–United States investigation of the role of context in

Table 2.1
Example teaching approaches for multiplying and dividing decimals (whole numbers and fractions) by 10, 100, 1000, and other multiples of 10

Approach	Teaching method	Resources	Assessment
1	Student research and collaborative knowledge building: what is the procedure and how/why does it work?	Internet (tutorials, games, YouTube, Google video); class wiki or blog for gradually building a class product to answer the question	Student self-assessment of learning and production of worked examples by individuals or pairs of students
2	Problem solving: group investigation of patterns in number sequences to discover the rule and creation of new patterns; teacher furnishes scaffolding as needed	Calculators	Teacher observation and group oral reports to the class
3	Teacher explanation of the procedure emphasizing place value, followed by written student exercises	Base-10 material and worksheets	Individual work samples

students' analysis of data, Nisbet, Langrall, and Mooney (2007) selected data sets (from the Internet) on two topics of interest to a group of middle-grades students and for one topic thought to be of little interest to them. The researchers devised tasks in the form of problems, with a view to stimulating higher-order thinking, and assigned students to small groups (fig. 2.2). Knowledge of context was an important factor contributing to students' engagement in statistical tasks, with students using context knowledge "to rationalize the data or their interpretations, in taking a critical stance toward the data" (Nisbet, Langrall, and Mooney 2007, p. 16). In the third task (based on Olympic discus-throwing distances), the topic lacked a class expert and drew minimal interest, so much less supportive discussion of relevant context factors took place. Embedding mathematical learning in real and relevant contexts, particularly contexts that let students use personal expertise, can increase their sense of control and self-worth.

Open-Ended Tasks

As well as learning to use mathematics in a range of contexts, students need to

Task 1: Does the USA produce better men tennis players than Australia?

Data sets: (*a*) a table listing information on the winners and runners-up for Wimbledon 1955–2004 and (*b*) a table of the top twenty male tennis players with rankings and prize money

Task 2: In a recent pop-culture survey, teenagers identified Britney Spears and Delta Goodrem as two of the top female performers. From the following data collected from the top 40 charts, which of these two singers is more popular?

Data sets: (*a*) a table for each singer of top 40 hits 2002–2005, including rankings and number of weeks, and (*b*) tables of top 40 ranking from charts around the world for the two singers

Fig. 2.2. Tasks to stimulate higher-order thinking (adapted from Nisbet, Langrall, and Mooney 2007)

develop deeper understanding for a range of important mathematical concepts (Kilpatrick, Swafford, and Findell 2001). Students also need to develop procedural fluency. To this end, teachers often use repetitive exercises to promote practice; overusing this approach may lead to student disengagement.

Another strategy that creates increased opportunities for motivation and engagement involves actively exploring mathematically rich situations through open-ended questions and tasks. Open-ended questions are more readily accessible since all students can start from their current knowledge and understanding. When planning to use open-ended questions, teachers need to design enabling prompts to support students who falter after only one or two solutions as well as additional prompts to extend thinking for students who complete the task more quickly (Sullivan, Mousley, and Zevenbergen 2006). Sharing problem-solving strategies in small-group situations can also enhance students' extended thinking.

Teachers can easily create open-ended questions by working backward from an answer or adapting a standard question (Small 2009; Sullivan and Lilburn 2004). The first approach involves selecting a topic and considering an answer to a typical question. Working backward from this answer, one can then derive open-ended questions. For example, when ordering fractions and giving students pairs of fractions to compare (e.g., "Which is bigger, $5/12$ or $3/5$?") a teacher may ask students to list some fractions less than $3/5$. By posing the additional challenge "Take one of your answers and explain how you know that it is less than $3/5$," teachers gain valuable assessment information about students' thinking and choice of strategies. If students find determining solutions difficult, some enabling prompts could include using a number line to locate $^3/_5$ or replacing $^3/_5$ with a simpler fraction such as $^1/_2$. A prompt to extend thinking would be to have students find fractions between $^2/_5$ and $^3/_5$.

The second approach involves adapting a standard question. For example, if students are learning about the properties of quadrilaterals, instead of using the question "What is a parallelogram?" the teacher could adapt the question by asking students to write down everything that they know about a parallelogram. Students having difficulty could begin by recording what they know about a rectangle and then comparing a rectangle with a parallelogram. The teacher can further engage students who easily complete the task by posing the additional challenge, "Compare and contrast the properties of a parallelogram with those of other quadrilaterals."

Open-ended questions may have many solutions and several methods of finding solutions. Such a format opens up opportunities for success, reducing anxiety and giving students more control as they decide what strategies to use. Open-ended questions also promote developing problem-solving skills and are generally considered less threatening than more challenging problem tasks. Instead of the "correct answer," the focus shifts to process, independence, flexibility, and persistence.

Using Errors as a Focus for Learning in Mathematics Lessons

Acknowledging the role of errors and misconceptions in the learning process also lets students take control of their learning and self-assess. Many factors may cause students' errors in mathematics. Poor comprehension, language difficulties, anxiety, rushing, and carelessness may lead to errors in completing tasks. However, misconceptions from overgeneralizing rules or failure to connect new ideas to existing knowledge and understanding usually cause systematic errors. Just as teachers review student scripts to look for patterns in errors, students need encouragement to review their errors, reinforcing the belief that making errors is an important part of learning. Changing beliefs such as "you cannot learn from your mistakes" will be necessary if students are to overcome failure avoidance, anxiety, and self-handicapping in mathematics.

Students' early mathematics learning may lead them to make generalizations such as "multiplication makes bigger," "division involves dividing a bigger number by a smaller number," and "longer numbers are always greater in value." Although these generalizations may be satisfactory with whole numbers in the early elementary grades, they may not apply for fractions and decimals. Teachers need to explicitly discuss such statements so that students are more aware of common misconceptions. Teachers could present one of these statements to students at the beginning of a lesson to encourage them to seek counterexamples.

A related teaching strategy is to pose a statement and ask students whether it is always, sometimes, or never true (Bills et al. 2004). Students discuss the possibilities in small groups, identifying examples and counterexamples to support their arguments. Further examples of statements that present opportunities for rich discussion include the following:

- Parallelograms have no axes of symmetry.
- To multiply a number by 10, you put a zero on the end.
- The larger the area of a shape, the larger the perimeter.
- All four-sided shapes tessellate.

Conclusion

Many teaching strategies that increase student motivation also positively influence academic achievement. Because both achievement and motivation are prone to falter in the crucial middle grades, educators must increase their understanding of these dimensions and their relationship to effective teaching strategies. In this chapter, we presented a use-inspired multidimensional framework for understanding motivation with direct applicability to teachers, parents, and students. The Motivation and Engagement Wheel can offer direction for mathematics instruction at the whole-class level and for developing targeted interventions for enhancing individual students' motivation. We have also suggested possible directions for what whole-class instructional approaches and targeted intervention strategies might look like through four examples: variety in teaching approaches, real and relevant tasks, open-ended questions, and using errors as a focus for learning. Building students' motivation in mathematics to enhance mathematics achievement, and as a vital end in itself, is an essential component of mathematics instruction.

REFERENCES

Australian Education Council. *A National Statement on Mathematics for Australian Schools*. Carlton, Victoria, Aus.: Curriculum Corp., 1990.

Bandura, Albert. *Self-Efficacy: The Exercise of Control*. New York: Freeman and Co., 1997.

Bills, Chris, Liz Bills, Anne Watson, and John Mason. *Thinkers—A Collection of Activities to Provoke Mathematical Thinking*. Derby, U.K.: Association of Teachers of Mathematics, 2004.

Covington, Martin V. *Making the Grade: A Self-Worth Perspective on Motivation and School Reform.* New York: Cambridge University Press, 1992.

Dweck, Carol S. "Motivational Processes Affecting Learning." *American Psychologist* 41 (October 1986): 1040–48.

Hill, Peter, Philip Holmes-Smith, and Kenneth Rowe. *School and Teacher Effectiveness in Victoria: Key Findings from Phase 1 of the Victorian Quality Schools Project.* Melbourne, Victoria, Aus.: Centre for Applied Educational Research, 1993.

Holmes, Emma. "Motivation: An Essential Component of Mathematics Instruction." In *Teaching and Learning Mathematics in the 1990s.* 1990 Yearbook of the National Council of Teachers of Mathematics (NCTM), edited by Tom Cooney and Christian Hirsch, pp. 101–7. Reston, Va.: NCTM, 1990.

Kilpatrick, Jeremy, Jane Swafford, and Bradford Findell, eds. *Adding It Up: Helping Children Learn Mathematics.* Washington, D.C.: National Academies Press, 2001.

Martin, Andrew J. "Examining a Multidimensional Model of Student Motivation and Engagement Using a Construct Validation Approach." *British Journal of Educational Psychology* 77 (June 2007): 413–40.

———. *The Motivation and Engagement Scale.* Sydney: Lifelong Achievement Group, 2008.

———. *Building Classroom Success: Eliminating Academic Fear and Failure.* London: Continuum, 2010.

Murphy, Karen, and Patricia Alexander. "A Motivated Exploration of Motivation Terminology." *Contemporary Educational Psychology* 25 (January 2000): 3–53.

National Council of Teachers of Mathematics (NCTM). *Principles and Standards for School Mathematics.* Reston, Va.: NCTM, 2000.

Nisbet, Steven, Cynthia Langrall, and Edward Mooney. "The Role of Context in Students' Analysis of Data." *Australian Primary Mathematics Classroom* 12, no. 1 (2007): 16–22.

Pintrich, Paul R. "A Motivational Science Perspective on the Role of Student Motivation in Learning and Teaching Contexts." *Journal of Educational Psychology* 95 (December 2003): 667–86.

Small, Marian. *Good Questions: Great Ways to Differentiate Instruction.* New York: Teachers College Press, National Council of Teachers of Mathematics, and Nelson Education, 2009.

Stipek, Deborah, Julie M. Salmon, Karen B. Givvin, Elham Kazemi, Geoffrey Saxe, and Valanne L. MacGyvers. "The Value (and Convergence) of Practices Suggested by Motivation Research and Promoted by Mathematics Education Reformers." *Journal for Research in Mathematics Education* 29 (July 1998): 465–88.

Stokes, Donald. *Pasteur's Quadrant: Basic Science and Technological Innovation.* Washington, D.C.: Brookings Institution Press, 1997.

Sullivan, Peter, and Pat Lilburn. *Open-Ended Maths Activities: Using "Good" Questions to Enhance Learning.* 2nd ed. Melbourne, Victoria, Aus.: Oxford University Press, 2004.

Sullivan, Peter, Judith Mousley, and Robyn Zevenbergen. "Teacher Actions to Maximize Mathematics Learning Opportunities in Heterogeneous Classrooms." *International Journal of Science and Mathematics Education* 4 (March 2006): 117–43.

Weiner, Bernard. "Principles for a Theory of Student Motivation and Their Application within an Attributional Framework." In *Research on Motivation in Education: Vol. 1., Student Motivation*, edited by Russell Ames and Carole Ames, pp. 15–38. New York: Academic Press, 1984.

Wigfield, Alan, and Jacquelynne Eccles. "Expectancy–Value Theory of Achievement Motivation." *Contemporary Educational Psychology* 25 (January 2000): 68–81.

Wiliam, Dylan. "Keeping Learning on Track: Classroom Assessment and the Regulation of Learning." In *Second Handbook of Research on Mathematics Teaching and Learning,* edited by Frank K. Lester Jr., pp. 1053–98. Charlotte, N.C.: Information Age Publishing and National Council of Teachers of Mathematics, 2007.

Zimmerman, Barry. "Achieving Self-Regulation: The Trial and Triumph of Adolescence." In *Academic Motivation of Adolescents*, edited by Frank Pajares and Tim Urdan, pp. 1–26. Charlotte, N.C.: Information Age Publishing, 2002.

Chapter 3

Identity Development: Critical Component for Learning in Mathematics

Alan Zollman
M Cecil Smith
Patricia Reisdorf

I'm beginning to understand myself. But it would have been great to be able to understand myself when I was 20 rather than when I was 82.

—Dave Brubeck, American jazz pianist

WHY IS it that, in one class, we have some students who mathematically achieve, and yet other students in the same class with similar aptitude and background who do not? What elusive attribute is the difference between these students? We argue that this attribute is student identity development. As mathematics teachers, we often ask ourselves the following:

- How do I motivate reluctant learners?
- How do I get all my students involved in learning mathematics?
- Why are some students motivated and ready to learn, whereas others are not?

We cycle through best practices of mathematics standards in curriculum, instruction, and assessment, yet find ourselves searching when some students, for unknown reasons, opt out of achieving mathematics proficiency. Much has been written about the kinds of classroom conditions necessary to establish appropriate motivational and academic goals that contribute to mathematics achievement among students. However, little of this work has devoted attention to the significance of identity development to foster appropriate attitudes and development of close teacher–student relationships that contribute to academic achievement (Boaler 2000).

In this paper we define student identity development related to mathematics learning and achievement, briefly review relevant literature, and offer ideas for creating classroom environments concentrated on mathematics learning where identity work is a central focus. First, however, a vignette illustrates several elements necessary to support student identity development in mathematics.

A Classroom Vignette

Walking down the hall to the classroom, we hear a bunch of noise. Entering the room, we notice lots of student work, proudly displayed, including a problem (posed by a student, Cesar) on the board: "There are 36 students in the Environment Club, with 8 more boys than girls. How many girls are in the club?" Then we hear, "Would your group please solve classmate Cesar's problem, using our algebra from last week?"

At first we do not see the teacher, Ms. Juarez. The students, all eighth graders, are in small groups, discussing the problem. We observe Ms. Juarez sitting with one of the groups:

Ms. Juarez: So you figured out the answer by guessing and testing various numbers for boys and girls; could you use the algebra we did last week to solve the problem more efficiently?

Channell: Well, we could plot the number of boys and girls as x and y points.

Ms. Juarez: If you did this, what would you be plotting?

Erik: If we connected the points, it would be a line.

Ms. Juarez: What would this line be, and what would it represent?

Breanne: It would be a line that represents the total number of students, like boys plus girls equals 36.

Channell: Yes, x plus y equals 36 students.

Maya: I think I see—we could do the same thing for 8 more boys than girls.

Erik: What do you mean?

Maya: Well, eight plus the number of boys equals the number of girls.

Erik: No, the number of girls has to be fewer.

Maya: Yeah, it's the number of boys minus 8 is the number of girls.

Channell: You mean, 'x minus 8 equals y.'

Maya: [*grinning*] Yes, that's how we should say it—we are mathematicians, after all!

Ms. Juarez: Okay, now what?

Channell: Well, we have two equations with the same variables. We could solve them.

Ms. Juarez: Are you sure you want to do this?

Erik: Yes, I just did, and it gives our answer.

Ms. Juarez: Erik, how did you solve them?

Erik: I put one equation into the other one.

Breanne: Wait, we did this another way last week. Remember?

Channell: Yeah, we can draw two lines; where they meet is the answer for both.

Ms. Juarez: This seems like a good method; what do you want to call it? Let's see what the other groups think of your ideas. Would you write this down to show other groups later? As usual, please create an additional problem, related to your own lives, that we can post for the whole class to work. Also, don't forget, your portfolios are due on Friday!

From this brief interaction, we see evidence of how Ms. Juarez initiates and supports identity formation of her mathematics students. The classroom is a collaborative community of learners (Lave and Wenger 1991; National Council of Teachers of Mathematics [NCTM] 1989, 1991, 2000). The cooperative learning group is engaged in an authentic problem, thinking aloud about what they know, and trying to connect previous ideas to the current topic. There is warmth to the interactions between Ms. Juarez and her students and mutual respect. She trusts her students, and they trust her. She grants them autonomy, the ability to make choices that determine what and how they will approach the mathematics problem, and she reinforces their developing sense of competence. Ms. Juarez expects and creates appropriate conditions for student persistence, respect, team membership, and integrity. The students understand their roles and responsibilities to the teacher and to themselves and their peers. We now define identity development and give a synopsis of literature relevant to mathematics learning.

Identity Development and Mathematics Learning

Cognitive, social, physical, and identity development are crucial to a student's mathematics achievement. According to Nakkula (2008), identity is the embodiment of self-understanding, and identity formation is the fundamental development task of psychological maturity during the adolescent years. Erikson (1968) describes eight stages of psychosocial development, from infancy to mature adulthood, each comprising a psychosocial crisis to be resolved and each building on the strengths (i.e., successful resolution of tasks) of the preceding stages.

Table 3.1 presents Erikson's psychosocial stage model. The task of identity formation is situated at the fifth stage. An identity crisis, which must resolve for healthy emotional and personality development to result in adulthood, precipitates identity formation. Identity resolution requires incorporating past and present identifications with significant others, developing and recognizing one's aptitudes and skills, defining one's values, and setting and working toward occupational goals and aspirations—all leading to a unified, integrated sense of self.

Table 3.1
Erikson's psychosocial stages

Stage	**Age level**	**Successful task resolution**	**Outcome**
Trust vs. mistrust	Infant	Infant develops trust when caregivers provide consistent care and affection	Hope
Autonomy vs. doubt	Toddler	Toddler works to master physical environment and be more independent of caregivers	Will
Initiative vs. guilt	Preschool	Preschooler asserts control and power over environment	Purpose
Industry vs. inferiority	School age	Child copes with new and increasing social and academic demands	Competence
Identity vs. role diffusion	Adolescence	Adolescent develops a sense of self and a personal identity	Fidelity

Table 3.1—*Continued*

Stage	Age level	Successful task resolution	Outcome
Intimacy vs. isolation	Young adult-hood	Adult forms intimate, loving relationships with others	Love
Generativity vs. stagnation	Middle adult-hood	Adult nurtures and supports the next generation	Care
Ego integrity vs. despair	Mature adult	Adult reflects on life with sense of fulfillment	Wisdom

All affective domains—motivation, self-esteem, self-confidence, beliefs, and attitudes—are associated with personal identity strivings. Students initiate identity work as they begin to think about their competencies and attributes, set short- and long-term goals, and evaluate personal beliefs. Adolescence is a period when students find and distinguish themselves from parents and family. Schools and peers afford important social contexts where much identity work occurs. Thus, the adolescent student must develop a sense of self. For mathematics, this means that students develop a view of themselves as capable, although not necessarily expert, mathematicians.

Identity work is an essential variable for many students' achievement. Students relate rejection of mathematics to the type of person that they believe themselves to be (Boaler, Wiliam, and Zevenbergen 2000). A student must first self-identify as a "mathematics student." Such an identity derives from the student's successes in the mathematics classroom, developing feelings of competence in mathematics, and becoming a valued member of a mathematics community of practice (Wenger 1998). The mathematics teacher therefore must create the appropriate classroom conditions for this identity work to flourish.

Identity Work in the Mathematics Classroom

Classroom identity work comprises four broad classes of teacher actions. These actions include both nurturing student needs for self-determination and furnishing instruction in self-regulating behavior. Further, these actions explicitly acknowledge the significance of student social worlds and their desires to interact with their peers. Establishing an engaging classroom environment that promotes student achievement also is essential to identity work. Table 3.2 summarizes the four classes of teacher actions that support student identity development, presenting also components and examples from the classroom.

Table 3.2
Four teaching actions that support students' identity development

Teaching action	Components	Classroom examples
Fostering self-determination	Autonomy, competence, relatedness	Give students choices; offer opportunities for practice and give informative, reinforcing feedback; create a classroom community emphasizing student interactions and shared learning
Cultivating self-regulation	Goal setting, selecting strategy, monitoring and evaluating effort, redirecting behavior	Teach specific learning strategies and problem-solving skills; codetermine student roles and responsibilities; require regular student reflection on actions and results of learning
Capitalizing on social goals	Membership in classroom community, cooperative learning	Establish a positive classroom community, warm and welcoming with mutual respect; expect team membership and integrity in cooperative learning groups
Establishing an engaging classroom environment	Emphasizing effort to improve skills and knowledge, developing positive self-efficacy, valuing learning, focusing on mastery, minimizing competition	Remind students to work hard and persist; use portfolios for assessment and self-reflection; use authentic problems; connect new topics to previous concepts; individualize instruction as needed to meet learners' needs

Fostering Self-Determination

Mathematics teachers must foster students' needs to be self-determining individuals. According to self-determination theory (Deci and Ryan 1985;), three basic human needs require nurturing for people to be psychologically healthy. First, students need to feel autonomous, that is, to have control and agency in their lives and actions. Second, they must become competent in the activities in which they engage. Finally, they need emotional connection to and support from others, as well as the ability to participate in mutually satisfying relationships. Deci and Ryan (1991) argue that meeting these needs is crucial not only to overall psychological well-being but also for sustaining student interest and commitment to academic learning tasks. To ensure that students develop a sense of personal agency and feel in control of their learning, teachers can best meet student self-determination needs by offering choices among activities, assignments, classroom tasks, or with whom to work. Authentic problem-solving situations that students value, and can connect to their personal backgrounds and experiences, contribute to their sense of competence and personal identity. In our classroom vignette, Ms. Juarez respected student autonomy regarding their approaches to Cesar's problem and through her question prompts.

Cultivating Self-Regulation

A second class of teacher actions focuses on giving learners instruction so as to cultivate self-regulation behaviors and strategies (e.g., goal setting, planning, contingency management), enabling academic success in mathematics (Pape, Bell, and Yetkin 2003), and contributing to feelings of competence (Deci and Ryan 1985). Cultivating self-regulation includes explicit instruction in self-regulation skills and strategies and tending to students' positive possible selves (Markus and Nurius 1986).

Self-regulation skills and strategies for learning help students learn how to identify and set goals, select among a repertoire of learning strategies, actively monitor and evaluate their efforts at goal attainment, and redirect their behaviors when they fall short of a goal. Such instruction encourages students to take control of their learning, an important ingredient in self-determination and appraisal of one's competence. Mathematics students should learn to reflect on what they are learning, how they learn mathematics best, and how to self-assess their progress in mathematics learning. Our teacher, Ms. Juarez, cultivated self-regulation behaviors and strategies by challenging students: "Let's see what the other groups think of your ideas. Would you write this down to show other groups later?"

Along with explicit instruction in self-regulation skills and strategies, attending to students' positive possible selves (Markus and Nurius 1986) is vital to identity work. Possible selves are one's ideas about what one wants to become (hoped-for self), wants to avoid becoming (feared self), and realistically expects to become (expected self). These beliefs can be highly motivating, according to Markus and Nurius, and are instrumental in guiding student self-regulation of their behaviors (Oyserman and Markus 1990; Ruvolo and Markus 1992), such as when studying mathematics. For example, when individuals have clear ideas about what they want to become (good mathematics students), they are more willing to put forth the necessary effort to attain their goals.

Possible selves furnish valuable reference points where students can evaluate their progress toward goals: are they engaging in behaviors that will let them achieve their hoped-for selves and avoid their feared selves (Cross and Markus 1994)? Also, a hoped-for self that is concrete, realistic, and detailed—and that invokes the necessary strategies for achieving a desired goal—will influence student behavior, producing the intended results over time (Oyserman and Markus

1990), thereby serving a self-regulatory role for learners. As Maya exclaimed in the vignette, "We are mathematicians, after all!"

Capitalizing on Social Goals

The third class of teacher actions is for teachers to capitalize on, rather than resist, students' social goals, including students' desires to interact with and get reinforcement from their peer group (Wentzel 1998). McCollum (2006) linked social goals to motivation and academic achievement, and social goals are cognitive representations of the social outcomes (e.g., friendship, care, feedback, status) that students desire.

Schools and classrooms are rich social environments, yet teachers often overlook student needs for socialization and peer interaction in favor of promoting academic work and success. Students spend much of their time in school socializing with friends and classmates and observing and learning from one another. Students who appear more interested in talking with their friends than in attending to mathematics can distract teachers, but only if such talking is not linked to motivation and academic social goals.

Most students want to feel that they belong (i.e., have a sense of membership) and want their school and classroom communities to welcome and acknowledge them as valued members (Faircloth 2009). A sense of belonging is related to student academic achievement (Cohen and Garcia 2008). One effective instructional method to increase student belongingness and academic performance in mathematics is cooperative learning (Slavin 1991). When students work in cooperative groups to learn mathematics, such work enhances their learning outcomes and feelings of competence and self-esteem, as Ms. Juarez's students illustrate. Cooperative learning also can expose students to individuals of diverse ethnic, racial, socioeconomic, and cultural backgrounds, thereby improving relationships across diverse groups. Teachers can implement cooperative learning groups in several ways, but their design should include an essential responsibility and individual accountability for each student.

Establishing an Engaging Classroom Environment

A fourth class of teacher actions is for teachers to establish an engaging classroom environment that orients students to strive for improvement and mastery of mathematics content and that emphasizes effort, improvement, and learning over competition and grades. The advantage of such classroom environments is that, as Roeser and Lau (2002) note, students are more likely to adopt mastery learning goals, develop positive beliefs about their academic efficacy, value and enjoy learning the subject matter more deeply, and use more effective learning strategies. To establish an engaging environment, mathematics teachers must create relevant and meaningful learning tasks (Zollman et al. 2009). Ms. Juarez understands the importance of regularly connecting mathematics to students' everyday lives through the challenges that she assigns. She encourages students to bring mathematics problems to the classroom that they encounter outside school.

Ms. Juarez creates a mathematics environment from the first day of class. She establishes her cooperative group rules. Students are responsible for their own learning, each student feels an individual commitment to be an active participant of a problem-solving team, and each student is answerable for other students' learning.

Students also can keep portfolios to see the development of their understanding. A portfolio system includes regular intervals for students to receive individualized reactions from Ms. Juarez and to evaluate and reflect on their learning. Using student portfolios also can be an important method for students to take ownership of their mathematics learning. For example, portfolios can enable students to display their competence, demonstrate their interest in particular areas of mathematics, and show their mathematics career aspirations (Maxwell and Lassak 2008).

Conclusion

The classroom teacher is the single most important in-school influence on student achievement and motivation (Darling-Hammond 1999). Teachers who are knowledgeable about, and sensitive to, identity development can best nurture adolescent student abilities, knowledge, and interests. How teachers establish and manage the classroom environment, and how they devise and employ academic tasks and activities, can promote identity development.

Mathematics teachers who embrace identity work as essential to instruction display distinct characteristics. Such teachers know and understand the importance of identity formation to mathematics achievement, and so they establish classroom environments conducive to student identity strivings. In doing so, they create classroom conditions that nurture student needs for self-determination by giving students opportunities to make choices, demonstrate their mathematics competency, and participate in supportive peer relationships. They teach, support, and encourage students' self-regulation behaviors, helping students evaluate their progress toward desired mathematics goals. Such teachers also encourage student exploration of positive possible selves to connect mathematics learning to considerations of their future lives. Rather than discouraging social interaction among students, mathematics teachers who engage in identity work expressly encourage and support student peer relationships and social goals. Using pedagogical practices such as cooperative learning lets students work and learn together. Finally, teachers emphasize student effort, improvement, and mastery of mathematics in order to help all learners become proficient in mathematics.

Many mathematics teachers worry about a student's physical, social, and intellectual development. Yet identity development is equally crucial for mathematics achievement. Self-development, including a strong sense of personal identity, is of course no guarantee that student motivation for academic achievement in mathematics will follow. Some students develop a strong understanding of their aptitudes and interests and can articulate well-chosen personal goals. Some will reject mathematics as uninteresting and irrelevant. For students who might not appreciate the beauty and significance of mathematics to everyday life (including career opportunities and intellectual engagement), identity development can be highly motivating for mathematics learning. Mathematics teachers also must model behaviors and attitudes of a lifelong, inquisitive learner (Plummer and Peterson 2009). Infusing lessons with passion for mathematics further stimulates student interest and enthusiasm and contributes to greater student engagement. We, as mathematics teachers, can offer many opportunities to encourage, support, and sustain student identity development in mathematics.

REFERENCES

Boaler, Jo. "Mathematics from Another World: Traditional Communities and the Alienation of Learners." *Journal of Mathematical Behavior* 18 (June 2000): 1–19.

Boaler, Jo, Dylan Wiliam, and Robyn Zevenbergen. "The Construction of Identity in Secondary Mathematics Education." Paper presented at the Second Mathematics Education and Society Conference, Montechoro, Algarve, Portugal, March 2000.

Cohen, Geoffrey L., and Julio Garcia. "Identity, Belonging, and Achievement: A Model, Interventions, Implications." *Current Directions in Psychological Science* 17 (December 2008): 365–69.

Cross, Susan E., and Hazel Rose Markus. "Self-Schemas, Possible Selves, and Competent Performance." *Journal of Educational Psychology* 86 (September 1994): 423–38.

Darling-Hammond, Linda. *Teacher Quality and Student Achievement: A Review of State Policy Evidence*. Seattle: Center for the Study of Teaching and Policy, 1999.

Deci, Edward L., and Richard M. Ryan. *Intrinsic Motivation and Self-Determination in Human Behavior.* New York: Plenum, 1985.

———. "A Motivational Approach to Self: Integration in Personality." In *Nebraska Symposium on Motivation: Vol. 38, Perspectives on Motivation*, edited by Richard Dienstbier, pp. 237–88. Lincoln, Neb.: University of Nebraska Press, 1991.

Erikson, Erik H. *Identity: Youth and Crisis.* New York: Norton, 1968.

Faircloth, Beverly S. "Making the Most of Adolescence: Harnessing the Search for Identity to Understand Classroom Belonging." *Journal of Adolescent Research* 24 (May 2009): 321–48.

Lave, Jean, and Etienne Wenger. *Situated Learning: Legitimate Peripheral Participation*. New York: Cambridge University Press, 1991.

Markus, Hazel, and Paula Nurius. "Possible Selves." *American Psychologist* 41 (September 1986): 954–69.

Maxwell, Vicki L., and Marshall B. Lassak. "Using an Experiment in Portfolios in the Middle School." *Mathematics Teaching in the Middle School* 13 (March 2008): 404–9.

McCollum, Daniel L. "Students' Social Goals and Outcomes." *Academic Exchange Quarterly* 10 (June 2006): 17–21.

Nakkula, M. "Identity and Possibility: Adolescent Development and the Potential of Schools." In *Adolescents at School: Perspectives on Youth, Identity, and Education*, edited by Michael Sadowski, pp. 11–22. Cambridge, Mass.: Harvard Education Press, 2008.

National Council of Teachers of Mathematics (NCTM). *Curriculum and Evaluation Standards for School Mathematics.* Reston, Va.: NCTM, 1989.

———. *Professional Standards for Teaching Mathematics.* Reston, Va.: NCTM, 1991.

———. *Principles and Standards for School Mathematics*. Reston, Va.: NCTM, 2000.

Oyserman, Daphna, and Hazel Markus. "Possible Selves in Balance: Implications for Delinquency." *Journal of Social Issues* 46 (Summer 1990): 141–57.

Pape, Stephen J., Clare V. Bell, and Iffet Elif Yetkin. "Developing Mathematical Thinking and Self-Regulated Learning: A Teaching Experiment in a Seventh-Grade Mathematics Classroom." *Educational Studies in Mathematics* 53, no. 3 (2003): 179–202.

Plummer, Julie Stafford, and Blake E. Peterson. "A Preservice Secondary Teacher's Moves to Protect Her View of Herself as a Mathematics Expert." *School Science and Mathematics* 109 (May 2009): 247–57.

Roeser, Robert W., and Shun Lau. "On Academic Identity Formations in Middle School Settings during Early Adolescence: A Motivational–Contextual Perspective." In *Understanding Early Adolescent Self and Identity: Applications and Interventions*, edited by Thomas M. Brinthaupt and Richard P. Lipka, pp. 91–131. Albany, N.Y.: State University of New York Press, 2002.

Ruvolo, Ann P., and Hazel Markus. "Possible Selves and Performance: The Power of Self-Relevant Imagery." *Social Cognition* 10 (Spring 1992): 95–124.

Slavin, Robert E. *Student Team Learning: A Practical Guide to Cooperative Learning.* 3rd ed. Washington, D.C.: National Education Association, 1991.

Wenger, Etienne. *Communities of Practice: Learning, Meaning, and Identity*. New York: Cambridge University Press, 1998.

Wentzel, Kathryn R. "Social Relationships and Motivation in Middle School: The Role of Parents, Teachers, and Peers." *Journal of Educational Psychology* 90 (June 1998): 202–9.

Zollman, Alan, Kathleen Kitts, Mansour Tahernezhadi, and Penny Billman. "Synergy For Science Learning: An Interdisciplinary Partnership to Improve the Quality of Science, Technology, Engineering, and Mathematics Education." *International Journal of Science in Society* 1 (December 2009): 147–52.

Chapter 4

Recommendations from Self-Determination Theory for Enhancing Motivation for Mathematics

Daniel J. Ross
David A. Bergin

In 2000, the National Council of Teachers of Mathematics (NCTM) released the *Principles and Standards for School Mathematics,* which offered recommendations for teaching K–12 mathematics. This chapter examines these suggestions in light of self-determination theory (SDT), a prominent theory of motivation that has been applied to classrooms. Consider the following vignette:

Ms. Whittier has been teaching algebra 1 for several years. She recognizes functions as one of the most important concepts in the course. She wants her students to be able to interpret graphs of functions and decide what information the graphs yield. In the past, she presented the activity in figure 4.1 to her students to work on in preassigned groups.

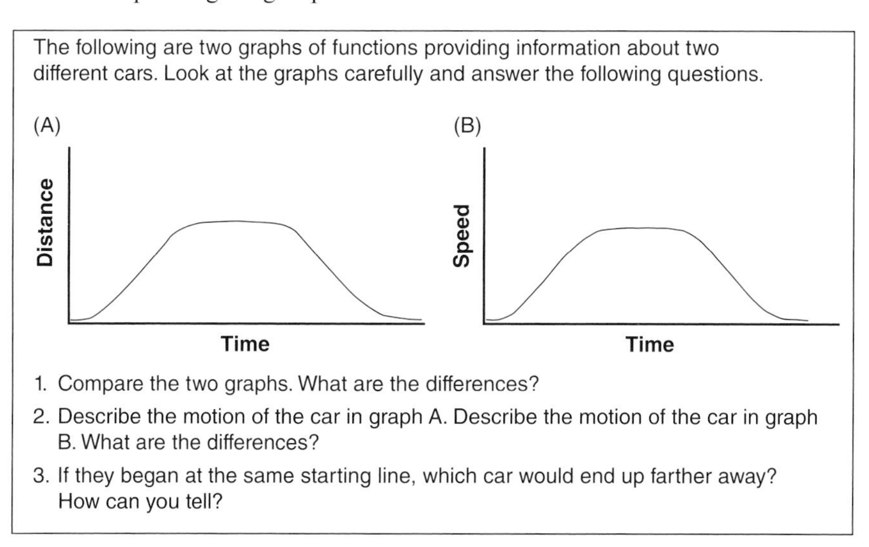

Fig. 4.1. Ms. Whittier's original activity about interpreting graphs of function

After the students had about twenty-five minutes to work on the questions, she had them turn in written responses for a grade. Typically, Ms. Whittier had some students who engaged with the questions and improved in interpreting graphical information about functions. However, she always had some students who did not engage in the problem or learn much from the activity.

This year, she decided to take a different approach in hopes of encouraging more students to engage and learn from the experience. First, she presented the graphs from figure 4.1 to the class. Then, rather than telling them that the graphs are about cars, she asked the students to take a few moments to examine the graphs and then describe possible scenarios that the graphs might depict. She recorded the suggestions on the board. She then highlighted three student-generated scenarios that the graphs could legitimately represent and told the students that they could choose one for their work on the problem. She explained that they would be working with one or two other students of their choice on making sense of the graphs and trying to understand and communicate to the rest of the class the differences between the functions. Using the scenario of their choice, they were to do one of the activities in figure 4.2.

Option 1	**Option 2**	**Option 3**
Write a story for each graph. Each story should be coherent and make sense. The story should include details that reveal what the graph represents and what the differences are between graphs.	Create an appropriate time–speed graph for the object in graph A. Create an appropriate time–distance graph for the object in graph B. Explain the further insight that one can gain from these new graphs and why they are different.	Re-create each situation (live or with a model) and video record the re-creation. Explain what is happening in each video and how this corresponds to the graph. Explain the differences between each video.

Fig. 4.2. Three options for function activities

Ms. Whittier knew that the activities were rather involved and planned to give feedback to each group as they worked on their projects. To support students' progress and understanding, she planned to have groups working on the same activity share ideas at several points in the process. When they finished their work, she asked them to share their results with the class. She also planned to post students' work on the school Web site and to include a notice in the school newsletter so that parents and others could find out what the students had done.

Did the changes that Ms. Whittier made improve her students' motivation to learn the mathematics that she identified as important? How did her new approach align with NCTM recommendations?

To answer these questions, let us take a closer look at the *Principles and Standards* (NCTM 2000) and a specific theory of motivation. *Principles and Standards* includes comments that imply a stance on motivating students in mathematics classrooms, though it does not discuss motivation explicitly. Many of the Teaching and Learning Principles, in particular, relate to motivating students. SDT gives us a useful framework for analyzing the *Principles and Standards*. This article shows how SDT aligns with the recommendations in the *Principles and Standards* and suggests specific practices that would support students' productive disposition toward, and their motivation to learn and achieve in, mathematics. Although much SDT research has occurred in classrooms, little has been specific to mathematics classrooms.

Self-Determination Theory's Three Fundamental Psychological Needs

SDT proposes three fundamental psychological needs innate to all humans: autonomy, competence, and relatedness. Each person's psychological well-being and growth depend on how fully these needs are satisfied. People are naturally inclined to pursue fulfilling these needs. When needs are not satisfied, people tend to experience depressed motivation, psychological problems, and ill-being.

Autonomy

Deci and Ryan (2000) described autonomy as the inner approval of one's actions and the sense that one's actions emanate from oneself. They pointed out that the concept does not refer to a sense of detachment or independence. For example, a student does not need to work on a different task from other students to feel autonomy. Rather, students' feeling that they chose to engage in the task for themselves supports their autonomy. When students feel that their behavior is autonomous, they feel an internal perceived locus of causality and experience self-determination. Thus, students whose need for autonomy is being met feel that their actions are under their own willful control.

Reeve and Jang (2006) determined eight teacher behaviors that support students' feelings of autonomy: increased time listening, increased time allowing students to work in their own way, increased time for student talk, praise as informational feedback, offering encouragement, offering hints, being responsive to student-generated questions, and making perspective-acknowledging statements. In the vignette, Ms. Whittier's changes have the potential to support students' feelings of autonomy because she increased the time that she listened to students talking, responded more to their input, offered more encouragement and feedback, and gave students more opportunities to work how they preferred.

In contrast to supporting autonomy, a controlling teacher who requires students to act contrary to their desires or goals does not meet students' needs for autonomy. Controlling-teacher behaviors include the following: not letting students handle learning materials, prematurely giving solutions, giving excessive commands, making "should" or "ought to" statements, asking controlling questions (e.g., "Can you do it the way I showed you?"), not allowing students to work at their own pace, and accepting only certain opinions and views (Assor et al. 2005; Reeve 2009). Reeve (2009) suggests that teachers are often overly controlling because they think that their role as teacher requires them to be in control and because the burdens of responsibility and accountability pressure them. They do not understand that excessive teacher control undermines student motivation.

Other factors such as rewards, threats, and deadlines can create a controlling environment that undermines autonomy (Reeve 2009). Deadlines and evaluation are inherent aspects of school settings, but teachers can present them in either highly controlling ways or, preferably, more matter-of-fact and informational ways. Extrinsic rewards, in particular, harm intrinsic motivation when high interest already exists. Extrinsic rewards cannot do much harm when no motivation exists, but SDT proponents generally prefer to avoid using them (Deci, Koestner, and Ryan 1999). Teachers can face tough decisions when deciding whether to use rewards. Teachers may have some students who experience interest and for whom rewards could harm developing a positive disposition, whereas they may have other students who lack interest and whom rewards could help. When rewards foster increased engagement and practice, students often develop increased perceptions of competence and motivation.

Consider Ms. Whittier's autonomy-related changes in the vignette. She exchanged an extrinsic reward (grade on homework) for the reward of making her students' accomplishments available on the Internet to share with people the students care about. She also reduced controlling behavior by giving students more opportunities to handle learning materials, letting them work at different paces, and allowing for more viewpoints.

Environments and events that support autonomy are generally associated with greater intrinsic motivation, greater interest and engagement, less pressure and tension, more creativity, greater excitement and vitality, more cognitive flexibility, improved conceptual learning and understanding, more positive emotionality, increased self-esteem and confidence, more trust, greater persistence, better physical and psychological well-being, and increased academic achievement (Deci and Ryan 2000; Reeve and Jang 2006; Ryan and Deci 2000). These positive outcomes suggest that teachers should seek ways to support students' autonomy.

Although *Principles and Standards* is careful not to promote a single way to teach, several recommendations align with factors that support autonomy, specifically, “A major goal of school mathematics programs is to create autonomous learners” (NCTM 2000, p. 21). *Principles and Standards* promotes student-centered instruction by encouraging teachers to center their teaching on learning about their students’ thinking, helping students build understanding, and actively engaging students in their learning. NCTM (2000) also suggests that teachers listen to students’ talk and respond to students’ questions. *Principles and Standards* also recommends that teachers give students choices and opportunities to work in their own way and emphasizes the importance of students’ conjecturing, reasoning, and arguing their mathematical thoughts as opposed to memorizing facts and procedures from the teacher. Finally, NCTM (2000) promotes using formative assessment to give students informative feedback. Together, these recommendations align with the preceding autonomy-supportive practices.

How could teachers support students’ autonomy in a mathematics classroom? As an example: secondary mathematics teachers who want their students to reason with and use trigonometric ratios of right triangles could give students three options, as in figure 4.3. Students would have the opportunity to choose tasks that interest them and still achieve the same learning goals.

Significantly, although increasing choices available to students generally supports autonomy, too many choices can be demotivating and lead to results of lower quality (Iyengar and Lepper 2000). Thus, mathematics teachers do not need to find ways to give students unlimited options at every opportunity. Instead, teachers should seek ways to create a few options that will be meaningful to students on regular occasions. For example, in the vignette, Ms. Whittier limited the students to three potential scenarios and three options for engaging with the problem. Allowing students to engage with the problem in any way that they wanted may have overwhelmed some and actually caused them to lose motivation. Figure 4.4 summarizes principles for supporting autonomy.

Competence

Deci and Ryan (2000) defined competence as the extent to which students feel that they can attain valued outcomes and explained how teachers can support feelings of competence. Events that elicit feelings of competence include challenges that are at just the right level, feedback that makes students feel successful, and freedom from evaluations that make students feel incompetent. When students engage in an activity that presents an optimal level of challenge, they believe that they can succeed with effort. Succeeding at a challenging activity meets their need for competence. However, if the activity is too challenging and they cannot succeed, or if the activity is too easy and success has little meaning, then it does not satisfy students’ need for competence.

Thus, mathematics teachers seeking to meet their students’ needs for competence need to furnish activities that are at optimal levels of challenge for each

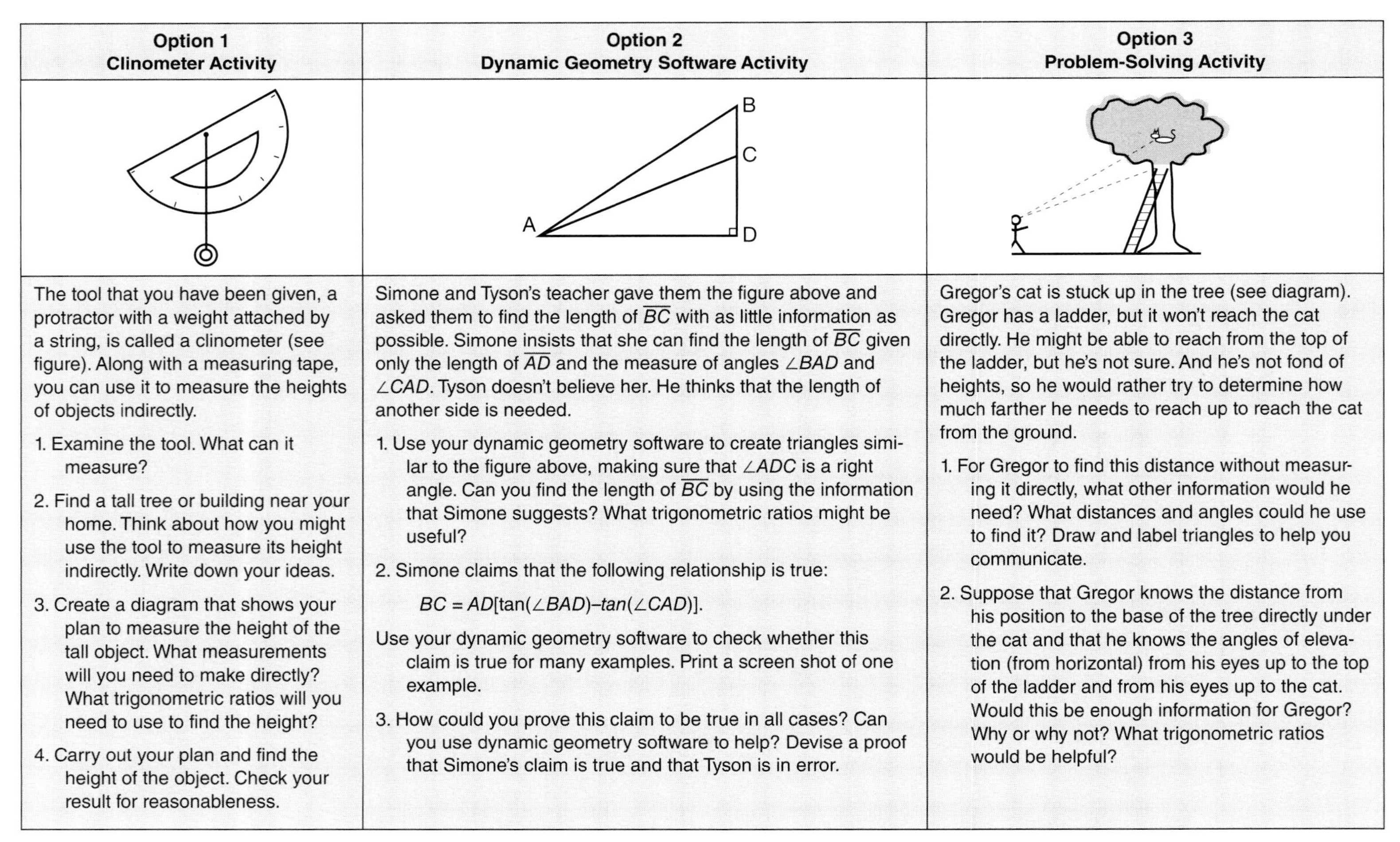

Option 1 Clinometer Activity	Option 2 Dynamic Geometry Software Activity	Option 3 Problem-Solving Activity
	A B C D	
The tool that you have been given, a protractor with a weight attached by a string, is called a clinometer (see figure). Along with a measuring tape, you can use it to measure the heights of objects indirectly. 1. Examine the tool. What can it measure? 2. Find a tall tree or building near your home. Think about how you might use the tool to measure its height indirectly. Write down your ideas. 3. Create a diagram that shows your plan to measure the height of the tall object. What measurements will you need to make directly? What trigonometric ratios will you need to use to find the height? 4. Carry out your plan and find the height of the object. Check your result for reasonableness.	Simone and Tyson's teacher gave them the figure above and asked them to find the length of $\overline{BC}$ with as little information as possible. Simone insists that she can find the length of $\overline{BC}$ given only the length of $\overline{AD}$ and the measure of angles $\angle BAD$ and $\angle CAD$. Tyson doesn't believe her. He thinks that the length of another side is needed. 1. Use your dynamic geometry software to create triangles similar to the figure above, making sure that $\angle ADC$ is a right angle. Can you find the length of $\overline{BC}$ by using the information that Simone suggests? What trigonometric ratios might be useful? 2. Simone claims that the following relationship is true: $BC = AD[\tan(\angle BAD) - tan(\angle CAD)]$. Use your dynamic geometry software to check whether this claim is true for many examples. Print a screen shot of one example. 3. How could you prove this claim to be true in all cases? Can you use dynamic geometry software to help? Devise a proof that Simone's claim is true and that Tyson is in error.	Gregor's cat is stuck up in the tree (see diagram). Gregor has a ladder, but it won't reach the cat directly. He might be able to reach from the top of the ladder, but he's not sure. And he's not fond of heights, so he would rather try to determine how much farther he needs to reach up to reach the cat from the ground. 1. For Gregor to find this distance without measuring it directly, what other information would he need? What distances and angles could he use to find it? Draw and label triangles to help you communicate. 2. Suppose that Gregor knows the distance from his position to the base of the tree directly under the cat and that he knows the angles of elevation (from horizontal) from his eyes up to the top of the ladder and from his eyes up to the cat. Would this be enough information for Gregor? Why or why not? What trigonometric ratios would be helpful?

Fig. 4.3. Three options for trigonometry activities

Do	Do Not
• Offer encouraging, informational feedback • Give meaningful rationale for tasks • Acknowledge students' perspectives • Give several choices • Listen and respond to students	• Use controlling language, rigid deadlines, and rewards • Prevent students from handling materials • Prematurely give solutions • Require students to work at a rigid pace • Accept only certain views

Fig. 4.4. Supporting autonomy

student. Students tend not to persist at tasks that are too difficult for them to succeed in. When tasks are too easy, students will feel little accomplishment if they complete the task. Such a situation does not support their sense of competence because they do not need to demonstrate competence to succeed at the task. Tasks should not only require sustained engagement and thinking at demanding levels but also allow students to succeed. Sustaining engagement and thinking at demanding levels is difficult for teachers because the optimal level of challenge may differ for each student in a classroom. In fact, requiring all students to engage with the same task in the same way will generally not meet their needs for competence. Optimal tasks often have multiple approaches, and teachers should present tasks of various difficulty levels when possible. In the vignette, Ms. Whittier changed the problem to increase the number of ways that students could approach the problem. Her changes also gave her opportunities to promote varied levels of challenge to meet the needs of different students. For example, if a group of students finished before other students, she could challenge them to create and interpret additional graphs, make conjectures, and generalize their findings.

Principles and Standards recommends that teachers facilitate a challenging learning environment and specifically notes that teachers should seek to provide challenging activities for students of various levels of ability (NCTM 2000, p. 19). NCTM argues for maintaining an appropriate level of difficulty so that students will engage with activities but will be required to work hard to succeed. One way to do so is to use tasks that allow multiple approaches so that students can use a variety of strategies to work on the activity, ranging from simple to more complex.

Principles and Standards also warns against teachers' reducing the challenge level of activities for students by "taking over the process of thinking" (p. 19). Hiebert and Grouws (2007) noted that one of the most consistent themes in research on teachers' effects on students' learning is that students' learning improves when teachers allow them to appropriately grapple with challenging activities. Unfortunately, a prevalent practice among U.S. mathematics teachers is reducing the challenge level before allowing students to attempt the activities (Hiebert et al. 2003). Consider a teacher who wants students to examine relationships between perimeter and area by posing the following question: "Farmer Tom has 100 meters of fencing to make a goat pen. What is the largest area the pen could be?" If the teacher immediately suggests that students try to find factors of 100 and shows them how to find the perimeter of rectangular pens with these factors, the teacher would be undermining students' opportunity to make sense of the problem. Also, according to SDT, reducing the cognitive demand of the problem prevents satisfying students' need for competence because such an approach does not let them demonstrate their competence on challenging activities. Recall from the previous section that such actions also tend to undermine perceptions of autonomy. When teachers continually reduce the challenge level of activities, students receive the message that they are not capable of successfully completing the activities.

Let us return to the example about relationships between perimeter and area. The teacher can support competence by enacting the task in ways that allow multiple approaches at different levels of difficulty. For example, some students could use a hands-on approach by drawing examples of goat pens on grid paper and comparing areas, whereas other students could seek to develop a function relating the dimensions of the pen with the resulting area and then finding the maximum value of the function. To support competence, the teacher should begin by giving students time to work individually on the task. Then, when students indicate that they are struggling, the teacher's goal should be to offer just enough support for students to continue to engage and make progress with the task; premature support can signal that the teacher thinks the student lacks ability. For example, the teacher could direct students who have created several examples of pens to make a table to track the dimensions and resulting areas and to look for patterns in the table. Another strategy could be for the teacher to pair up students who have written different functions and ask them to compare their work. For a greater challenge for students who determine the optimal dimensions for a rectangular pen, the teacher could ask them to consider whether pens with different shapes could have greater area. Finally, to support competence, the teacher should have multiple students share their thinking on the task with the whole class so that students have multiple opportunities to see how to approach and solve the problem. Figure 4.5 summarizes principles for supporting competence.

Relatedness

Relatedness is the desire to feel connections to others (Deci and Ryan 2000). Research supports the general assertion that "the need to belong is a powerful,

Do	Do Not
• Supply activities with optimal levels of challenge • Use flexible activities that address the same mathematical content but can have different levels of challenge • Give students choices so that they can find appropriately challenging tasks	• Use the same task for all students without regard for different optimal levels of challenge • Reduce levels of challenge by giving too much help too early before students have an opportunity to engage with the task

Fig. 4.5. Supporting competence

fundamental, and extremely pervasive motivation" (Baumeister and Leary 1995, p. 497). The relatedness need not occur in the moment of instruction but should occur in a general base of relatedness toward teachers and peers. In classroom settings, a teacher can work early in the school year to establish positive norms for classroom and small-group interactions that help students communicate respectfully with each other. Having established these norms, the teacher need not explicitly build relatedness aspects into every activity because the typical way that the students interact supports their relatedness.

In mathematics classrooms where students feel safe to work with others and teachers demonstrate genuine interest in and concern for students, teachers and students are likely to feel stronger bonds with each other. Although this interest can be in areas of the students' lives outside mathematics, teachers should also show genuine interest in their students' mathematical thinking. For example, rather than responding to a student's answer by indicating "correct" or "incorrect," a teacher can follow any student response by asking, "How do you know that?" However, students typically view such questions as a signal of incorrect answers; teachers need to retrain them to think otherwise. Teachers can promote relatedness among students by creating a secure classroom in which students can share their thinking with each other. When students feel safe to share their thoughts as well as consider and respond to other students' thinking, they will feel more connections to each other. Group work and cooperative learning foster relatedness by cultivating interpersonal warmth and support. In a study of several hundred grade 1–5 students, emotionally warm, sensitive teachers had students with greater growth in math and reading ability than other teachers (Pianta et al. 2008).

Bergin and Bergin (2009) suggested how teachers can increase bonding in the classroom:

1. Increase sensitivity and warm, positive interactions with students.
   Teacher sensitivity refers to accurate detection and interpretation of

children's cues, provision of comfort, and responsiveness to distress.

2. Be well prepared for class and hold high expectations for students. Doing so shows that teachers care about student achievement and fits with the *Principles and Standards*.
3. Respond to students' agendas by offering appropriate choice whenever possible.
4. Help students be kind, helpful, and accepting of one another. Supportive peer culture is an essential component of school bonding.
5. Use noncoercive discipline. Coercive discipline depends on threats; the teacher's superior power; and the teacher's ability to control resources such as recess time, grades, and detentions.
6. Repair relationships that have high levels of conflict and in which the teacher has been controlling and dominating with the student.

In the vignette, Ms. Whittier may have initially faced challenges implementing her changes had she not already built up students' sense of relatedness through establishing supportive classroom norms and practices. However, the new task let her develop her relationships with students as she listened to their input and gave encouraging feedback. Students could also feel greater relatedness with each other as they chose with whom they worked, shared across groups as they worked, and presented their work to the class in the end. Sharing their work outside the classroom could also have developed a sense of community.

NCTM envisions classrooms as supportive places for learning. For all students to have fair access to learning, teachers must create classrooms in which students are secure. Such classrooms are more likely to satisfy students' needs for relatedness. "To support classroom discourse effectively, teachers must build a community in which students will feel free to express their ideas" (NCTM 2000, p. 61). When students feel safe in sharing their ideas, they build connections to others in the classroom. Figure 4.6 lists principles for supporting relatedness.

Do	Do Not
• Establish norms early that support positive interactions in the classroom • Show genuine interest in students • Be well prepared and hold high expectations • Respond to students and offer choice • Work to repair damaged relationships	• Use coercive discipline • Criticize students • Ignore or give only minimal response to students

Fig. 4.6. Supporting relatedness

Extrinsic and Intrinsic Motivation

Intrinsic motivation is the tendency to seek out novelty and challenges for inherent satisfactions rather than some separable consequences (Deci and Ryan 2000). Proponents of SDT claim that fulfilling the fundamental needs of competence, autonomy, and relatedness fosters intrinsic motivation (Deci and Ryan 2000). SDT does not presume that motivation for activities is either wholly extrinsic or intrinsic but instead proposes a continuum of motivation ranging from amotivation through several levels of extrinsic motivation and on to intrinsic motivation (Ryan and Deci 2000). The vital distinction among the forms of motivation is the degree to which the behavior is self-determined (fig. 4.7).

Type of motivation	Amotivation	Extrinsic				Intrinsic
Type of regulation	Nonregulation	External	Introjected	Identified	Integrated	Intrinsic regulation
Level of self-determination	Non–self-determined	⟶				Self-determination

Fig. 4.7. Continuum of types of motivation, regulation, and self-determination (adapted from Ryan and Deci [2000], p. 72)

At the ideal end of the continuum is the state of intrinsic motivation. Intrinsically motivated students in mathematics classrooms engage with the material of their own accord. In contrast, students with identified and integrated regulation experience only some self-determination and autonomy. They respond to an external source of motivation, but they understand the underlying value of the task, which is instrumental in accomplishing another goal. Thus, a student in twelfth grade may not be intrinsically motivated to study calculus but may understand it to be instrumental in pursuing an engineering career. *Principles and Standards* appears to endorse this type of motivation when it emphasizes that students should recognize the value of mathematics in future careers.

NCTM's (2000) recommendations support intrinsic motivation to learn mathematics by arguing that mathematics teachers and courses be designed to build on students' natural desires to understand and learn. *Principles and Standards* emphasizes that students should be engaged and interested as they learn mathematics and recommends that teachers adapt instruction to students' mathematical goals and supply opportunities for them to be active in their learning.

Doing so will support students' engagement, interest, and curiosity. NCTM's ultimate vision is that students "value mathematics and engage actively in learning [mathematics]" (NCTM 2000, p. 3).

Movement between Types of Regulation

Proponents of SDT argue that when students' experiences consistently thwart their needs for autonomy, competence, and relatedness, they move toward amotivation, and students who are amotivated toward learning tasks will not learn the content embedded in those tasks well (Deci and Ryan 2000). When students feel a lack of choice or control over their actions related to a task, low competence in a task, and poor relatedness with the teacher and other students involved in the task, they will feel less ability to succeed at the task, less interest in the task, and less reason to pursue it. Thus, they become amotivated toward the task.

Deci and Ryan (2000) argued that a first step toward internalization can come through students' feelings of relatedness. Thus, students who feel that their teachers care about them are more likely to respond to mathematical tasks even without internal reasons to engage. Middle school students felt that their teachers cared for them if the teachers helped each student academically, such as asking whether they needed help, calling on them, making sure that they understood content, teaching in a special way, and making class interesting (Wentzel 1997). Noncaring teachers got off task, taught while students weren't paying attention, and did not answer questions or explain things. Thus, in the vignette, Ms. Whittier's attention to students' needs for relatedness would influence the potential for her changes to improve motivation in some students.

Individuals are likely to meet and express needs differently in cultures that hold different values. Thus, prescribing specific actions that would support autonomy, competence, and relatedness for all students is impossible. Teachers must consider their specific students and situation to determine how best to meet their students' needs. For example, Asian American elementary school students who spoke their native language at home were actually better motivated by having others with whom they had some relationship (e.g., mothers or classmates) choose tasks for them rather than by having individual free choice (Iyengar and Lepper 1999). For these students, their needs for autonomy may have taken a secondary place to their need for relatedness.

Conclusion

General agreement exists between the recommendations in *Principles and Standards* and those of SDT. This congruence affords broad support for the vision described in *Principles and Standards* and for teachers who pursue its guidelines in their classrooms.

Our analysis revealed several important points in regard to students' motivation in mathematics education. Teachers should support students' autonomy and intrinsic motivation by offering them choices. However, unlimited increase in choice does not necessarily increase motivation. Instead, teachers should create a few choices that allow students to sense that students' actions emanate from themselves. Teachers should also support students' competence and relatedness. However, cultural differences can influence how students' autonomy, competence, and relatedness needs are perceived and met. When teachers find appropriate ways to meet their students' needs, students will internalize their motivation more. With intrinsic motivation to learn mathematics, students experience learning in the manner that *Principles and Standards* envisions.

REFERENCES

Assor, Avi, Haya Kaplan, Yaniv Kanat-Maymon, and Guy Roth. "Directly Controlling Teacher Behaviors as Predictors of Poor Motivation and Engagement in Girls and Boys: The Role of Anger and Anxiety." *Learning and Instruction* 15 (October 2005): 397–413.

Baumeister, Roy F., and Mark R. Leary. "The Need to Belong: Desire for Interpersonal Attachments as a Fundamental Human Motivation." *Psychological Bulletin* 117 (May 1995): 497–529.

Bergin, Christi A., and David A. Bergin. "Attachment in the Classroom." *Educational Psychology Review* 21 (June 2009): 141–70.

Deci, Edward L., Richard Koestner, and Richard M. Ryan. "A Meta-Analytic Review of Experiments Examining the Effects of Extrinsic Rewards on Intrinsic Motivation." *Psychological Bulletin* 125 (November 1999): 627–68.

Deci, Edward L., and Richard M. Ryan. "The 'What' and 'Why' of Goal Pursuits: Human Needs and the Self-Determination of Behavior." *Psychological Inquiry* 11 (October 2000): 227–68.

Hiebert, James, Ronald Gallimore, Helen Garnier, Karen Bogard Givvin, Hilary Hollingsworth, Jennifer Jacobs, Angel Miu-Ying Chui, et al. *Teaching Mathematics in Seven Countries: Results from the TIMSS 1999 Video Study*. Washington, D.C.: U.S. Department of Education, 2003.

Hiebert, James, and Douglas A. Grouws. "The Effects of Classroom Mathematics Teaching on Students' Learning." In *Second Handbook of Research on Mathematics Teaching and Learning*, edited by Frank K. Lester Jr., p. 371–404. Charlotte, N.C.: Information Age Publishing and National Council of Teachers of Mathematics, 2007.

Iyengar, Sheena S., and Mark R. Lepper. "Rethinking the Value of Choice: A Cultural Perspective on Intrinsic Motivation." *Journal of Personality and Social Psychology* 76 (March 1999): 349–66.

———. "When Choice Is Demotivating: Can One Desire Too Much of a Good Thing?" *Journal of Personality and Social Psychology* 79 (December 2000): 995–1006.

National Council of Teachers of Mathematics (NCTM). *Principles and Standards for School Mathematics*. Reston, Va.: NCTM, 2000.

Pianta, Robert C., Jay Belsky, Nathan Vandergrift, Renate M. Houts, and Frederick J. Morrison. "Classroom Effects on Children's Achievement Trajectories in Elementary School." *American Educational Research Journal* 45 (June 2008): 365–97.

Reeve, Johnmarshall. "Why Teachers Adopt a Controlling Motivating Style toward Students and How They Can Become More Autonomy Supportive." *Educational Psychologist* 44 (July 2009): 159–75.

Reeve, Johnmarshall, and Hyungshim Jang. "What Teachers Say and Do to Support Students' Autonomy During a Learning Activity." *Journal of Educational Psychology* 98 (February 2006): 209–18.

Ryan, Richard M., and Edward L. Deci. "Self-Determination Theory and the Facilitation of Intrinsic Motivation, Social Development, and Well-Being." *American Psychologist* 55 (January 2000): 68–78.

Wentzel, Kathryn R. "Student Motivation in Middle School: The Role of Perceived Pedagogical Caring." *Journal of Educational Psychology* 89 (September 1997): 411–19.

Chapter 5

Student Dispositions with Respect to Mathematics: What Current Literature Says

James Beyers

CONSIDERABLE efforts of mathematics educators have focused on explaining how students learn: which approaches are effective, for whom, and why. In the last twenty years or so, students' mathematical dispositions have received considerable attention (e.g., National Council of Teachers of Mathematics [NCTM] 1989, 2000), as have their dispositions toward mathematics (e.g., Kilpatrick, Swafford, and Findell 2001) and how those dispositions can influence students' development of mathematical knowledge. The dispositions that students form can influence not only their mathematical thinking and performance in the moment but also their attitudes and decisions about mathematics in later years (NCTM 1989).

Features of students' dispositions with respect to mathematics—which include, but are not limited to, beliefs about the nature of mathematics, math anxiety, and persistence—also play a crucial role in shaping what is taught and learned (NCTM 1989, 2000; Kilpatrick, Swafford, and Findell 2001). Students' dispositions with respect to mathematics can influence how students may or may not take advantage of opportunities to learn mathematics (Kilpatrick, Swafford, and Findell 2001). Opportunities to learn mathematics are "circumstances that allow students to engage in and spend time on academic tasks such as working on problems, exploring situations and gathering data, listening to explanations, reading texts, or conjecturing and justifying" (Kilpatrick, Swafford, and Findell 2001, p. 333). Opportunity to learn is one of the most important predictors of student achievement (Berliner and Biddle 1995; Kilpatrick, Swafford, and Findell 2001). Thus, educators must take seriously the goal of instilling dispositions that do not detract from students' learning of mathematics.

Frameworks exist for thinking about mathematics knowledge, such as procedural or conceptual (Hiebert 1986) or relational and instrumental (Skemp 1976); however, helpful frameworks for thinking about other important issues in mathematics education, such as students' dispositions with respect to mathematics, do not always exist (Boaler 2002). The primary contribution of this work is to forward a framework, based on a review of the literature, for thinking about students' dispositions with respect to mathematics in light of dispositional cognitive, affective, and conative mental functions. (The conceptual framework section will define these terms.)

To better understand the nature of student dispositions, I reviewed the literature. The results suggest that (1) the literature uses varied and at times inconsistent conceptualizations of the disposition construct; (2) among several factors, teachers play an essential role in shaping students' dispositions with respect to mathematics; and (3) students' dispositions with respect to mathematics affect student learning by means of opportunities to learn.

Finding What Research Has Said about Dispositions

This review investigates research on students' dispositions with respect to mathematics, that is, their mathematical dispositions or dispositions toward mathematics. The review included studies that examined students' dispositions with respect to mathematics as the main, secondary, or minor focus of the investigation. The studies also offered definitions or operationalizations of the disposition construct that were identifiably consistent with the aim and scope in this paper's conceptual framework section.

I sorted results according to the three lines of inquiry that emerged in reviewing the articles: conceptualizations of the disposition construct, development of dispositions, and impact on learning mathematics. I coded each article for inclusion or exclusion in each category; however, this paper presents only the results related to the conceptualizations and operationalizations of the disposition construct.

A Framework for Making Sense of Dispositions: Three Categories of Mental Processes

This section outlines a conceptual framework taken from educational psychology coupled with relevant constructs and definitions in mathematics education. The synthesized definition that guided the review follows the overview.

Historically, research has classified mental processes into three modes of mental functioning: cognition, affect (also called affection), and conation (also called volition; Snow, Corno, and Jackson III 1996). NCTM (1989) suggested thinking of a mathematical disposition as not simply attitudes but a tendency to think and to act in positive ways (p. 233). Considering the three modes of mental functioning together with NCTM's recommendation, one might infer the existence of dispositional cognitive, affective, and conative mental functions that contribute to a student's mathematical disposition.

Cognitive mental functions are "process[es] whereby an organism becomes aware or obtains knowledge of an object . . . [Cognition] includes perceiving, recognizing, conceiving, judging, reasoning . . . in modern usage sensing is usually included under cognition" (English and English 1958, p. 92).

This review considers a cognitive mental function to be a dispositional cognitive function with respect to mathematics if a person has a tendency or inclination to engage (or not) in a particular cognitive mental process associated with perceiving, recognizing, conceiving, judging, reasoning, and so on, in mathematics. For example, according to Boaler (2002), perhaps some students could have extensive knowledge of multiple areas in mathematics but could simply not have a tendency to make any mathematical connections among those areas. Given that mathematical knowledge can derive from making mathematical connections (Noss, Healy, and Hoyles 1997) and that some students may not be inclined to make mathematical connections (Boaler 2002), one can reasonably think of the cognitive function of making mathematical connections as dispositional. Affective mental functions are "a class name for feeling, emotion, mood, temperament . . . a single feeling response to a particular object or idea . . . the general reaction to something liked or disliked . . . the dynamic or essential quality of an emotion; the energy of an emotion" (English and English 1958, p. 15). McLeod (1992) suggested that attitudes toward mathematics, beliefs about mathematics as well as about one's self (in relation to mathematics), and emotions (e.g., joy or aesthetic responses to mathematics) reside within the affective domain. An affective mental function is a dispositional affective function with respect to mathematics if a person has a tendency or inclination to have or experience particular attitudes, beliefs, feelings, emotions, moods, or temperaments with respect to mathematics. Alpert and Haber (1960) asserted that students who experience debilitative math anxiety avoid mathematical tasks and thus the perceived source of the anxiety. Such persons tend to experience angst when engaged in mathematical activity, and consequently one can think of the affective function of anxiety as dispositional.

Conative mental functions are "that aspect of mental process or behavior by which it tends to develop into something else; an intrinsic unrest of the organism . . . almost the opposite of homeostasis. A conscious tendency to act, a conscious striving . . . It is now seldom used as a specific form of behavior, rather for an aspect found in all. Impulse, desire, volition, [and] purposive striving all emphasize the conative aspect" (English and English 1958, p. 104). A conative mental function is a dispositional conative function if a person has a tendency or inclination to purposively strive or to exercise diligence, effort, or persistence in the face of mathematical activity. For example, Dweck (1986) argued that students who believe that ability is fixed in mathematics may place less value on effort in doing mathematics and consequently are more likely to be easily discouraged in the face of challenging mathematical tasks. Schoenfeld (1989) observed a belief among students that they should solve math problems in a relatively short time. This belief may lead students to place little value on persistence, and consequently, exert less effort in solving more challenging nonroutine tasks. Therefore, some students may tend to exhibit varied levels of persistence or effort and be less likely to purposefully strive in the face of challenging mathematical tasks, supporting the assertion that one can think of conative functions as dispositional.

On the basis of the previously described work, then, one can consider dispositions with respect to mathematics to be those cognitive, affective, and conative functions that a student of mathematics tends or is inclined to engage or espouse in a mathematical context (e.g., doing and/or learning mathematics). A student's disposition with respect to mathematics is, therefore, the composition of those dispositional cognitive, affective, and conative mental functions that he or she engenders with exposure to mathematics. The next section presents results from the review of literature related to these constructs.

Results: How Researchers Have Used the Disposition Construct

In exploring the relationship between students' dispositions and learning, researchers have put forth several conceptualizations and operationalizations of the disposition construct. Throughout the literature relevant to students' dispositions with respect to mathematics, researchers commonly use two umbrella terms: mathematical dispositions and dispositions toward mathematics. Among the uses of the term *disposition* are both commonalities and differences. More specifically, "tendency" or "inclination" is prevalent among the definitions of the disposition construct (e.g., Moldavan and Mullis 1998; NCTM 1989; Kilpatrick, Swafford, and Findell 2001). However, conceptualizations of the construct as "dispositions toward mathematics" focus almost exclusively on dispositional affective and conative functioning, whereas conceptualizations of "mathematical dispositions" included elements of dispositional cognitive functioning in addition to dispositional affective and conative functioning.

For example, Kilpatrick, Swafford, and Findell (2001), in the National Research Council's (NRC) discussion of students' dispositions toward mathematics, focused primarily on issues associated with affective distinctions of mental processes (e.g., mathematics self-concept and beliefs about mathematics) and conative distinctions (e.g., that effort rather than ability matters), with no attention to cognitive distinctions such as making mathematical arguments. See table 5.1.

Table 5.1
Categories and descriptions of dispositional functions

Dispositional function type	**Example descriptions of disposition construct**
Cognitive	Making mathematical arguments,[a] making mathematical connections[a]
Affective	Beliefs about oneself as a learner of mathematics,[b] attitude toward mathematics,[c] beliefs about the nature of mathematics,[b] beliefs about the usefulness of mathematics,[c] beliefs about whether learning mathematics is worthwhile,[b] beliefs about whether mathematics is sensible[b]
Conative	Effort, diligence, persistence[b]

[a]Associated primarily with the term *mathematical disposition*.
[b]Associated primarily with the term *disposition toward mathematics*.
[c]Associated with both terms.

Mathematical Disposition

Curriculum and Evaluation Standards for School Mathematics defined students' mathematical dispositions as "not simply attitudes but a tendency to think and to act in positive ways" (NCTM 1989, p. 233). This definition is broad, and consequently, it is open to a plethora of interpretations. Nonetheless, consistencies are evident between the NCTM (1989) definition and other qualifications of students' mathematical dispositions. For example, McIntosh (1997) outlined various means for conducting informal formative assessments of students' dispositions for improving mathematics instruction. To clarify what *students' dispositions* meant, McIntosh offered the following definition: "one's usual mood; temperament, a habitual inclination, tendency" (p. 93).

McIntosh (1997) further qualified students' dispositions by articulating components of a student's mathematical disposition, suggesting the following constituents of a mathematical disposition: attitudes, beliefs, persistence, confidence, cooperative skills, and rejection of stereotypes about mathematics. Consistencies exist between the NCTM (1989) definition and McIntosh's description of a mathematical disposition: both speak to students' attitudes and tendencies toward certain ways of thinking and doing.

Also, other researchers suggest different qualifications of students' mathematical dispositions. Royster, Harris, and Schoeps (1999), in a study of college students' mathematical dispositions, cited the definition in NCTM (1989), adding that confidence, perseverance, and interest also indicate students' mathematical dispositions. Similarly, McClain and Cobb (2001), in an analysis of the development of sociomathematical norms in a first-grade classroom, used NCTM's notion of students' mathematical dispositions; however, just as Royster and colleagues did, McClain and Cobb further qualified the conceptualization of mathematical disposition in the context of a first-grade mathematics classroom. For McClain and Cobb (2001), a mathematical disposition deemed to be consistent with what NCTM (1989) advocated meant that students should develop some degree of mathematical autonomy. Mathematical autonomy in this context referred to students' ability to distinguish mathematically acceptable explanations from unacceptable explanations and to understand why students' explanations or strategies were or were not mathematically different. Both the conceptualizations of Royster, Harris, and Schoeps (1999) and McClain and Cobb (2001) seem to contain elements related to positive ways of thinking and acting, as evidenced by their reference to NCTM's (1989) conceptualization of mathematical dispositions. However, in both studies, the researchers apparently felt the need to further qualify the scope of mathematical dispositions, albeit in substantially different ways.

Although both studies referred to the same construct, students' mathematical dispositions, markedly different operationalizations of the construct occurred. The Royster, Harris, and Schoeps (1999) conceptualization supplemented NCTM's (1989) definition with elements of confidence, perseverance, and interest; in contrast, McClain and Cobb (2001) interpreted NCTM's (1989) definition as tendencies to assess mathematical explanations and mathematical difference. These different operationalizations are understandable in one respect, since the populations of interest were different—first graders versus college mathematics students—but wondering whether the construct "mathematical disposition" should be so different across grade bands seems reasonable. In all but one use of mathematical dispositions discussed earlier, the preceding authors (Royster et al. and McClain and Cobb, but not McIntosh) cited NCTM (1989) and its definition of mathematical disposition; however, all the authors described in more detail how they were conceptualizing students' mathematical dispositions. Each conceptualization, to various extents, included elements of dispositional affective and conative functioning, but McClain and Cobb (2001) also included elements of dispositional cognitive functioning—that is, tendencies to assess mathematical explanations and mathematical difference, elements absent from conceptualizations of dispositions toward mathematics.

Dispositions toward Mathematics

Many researchers have conceptualized students' dispositions toward mathematics (e.g., Fernandez and Cannon 2005; Kilpatrick, Swafford, and Findell 2001). The main difference between conceptualizations of "mathematical dispositions" and "dispositions toward mathematics" seems to lie in the absence of attention to elements of dispositional cognitive functioning.

The NRC (Kilpatrick, Swafford, and Findell 2001) offers a definition of students' dispositions toward mathematics that does not attend to students' dispositional cognitive functioning. The NRC defines a productive disposition toward mathematics as a "habitual inclination to see mathematics as sensible, useful, and worthwhile, coupled with a belief in diligence and one's own efficacy" (p. 116). Such a disposition refers to "a tendency to see sense in mathematics, to perceive it as both useful and worthwhile, to believe that steady effort in learning mathematics pays off, and to see oneself as an effective learner of mathematics" (p. 131). The NRC definition attends only to elements of dispositional affective functioning (e.g., an emphasis of beliefs about mathematics and self-confidence) and conative functioning (e.g., diligence).

Similarly, Fernandez and Cannon (2005) operationalized students' dispositions toward mathematics in the context of constructs that do not attend to dispositional cognitive functions. Fernandez and Cannon investigated what U.S. and Japanese teachers consider when planning and constructing mathematics lessons. Fernandez and Cannon coded teachers' descriptions of planning goals in light of content (mathematical goals) versus a desire to promote productive dispositions toward mathematics (dispositional goals). The researchers then categorized the dispositional goals, and only two were prevalent (i.e., at least 10 percent of respondents): (1) helping children appreciate relationships between mathematics and everyday experiences and (2) increasing students' fun and enjoyment while engaged in mathematical activity. Although the authors do not explicitly describe their criteria for deciding what constitutes a dispositional goal, one can surmise from the discussion that they coded as dispositional any goals that attended to promoting students' motivation or attention, appreciation of the relevance of mathematics to everyday experiences, and the fun and enjoyment associated with mathematics. The primary focus of this qualification seems to be more closely associated with beliefs about mathematics (e.g., a belief that mathematics is related to everyday experiences) as well as attitudes toward mathematics (e.g., fun and enjoyment of mathematics), both elements of dispositional affective functioning.

The NRC definition is similar to the qualification of Fernandez and Cannon in that students' perceptions of mathematics as useful may be commensurate with students' appreciation of the relevance of mathematics to their everyday experiences, because seeing how mathematics is useful in everyday experiences may be one way to appreciate the relationship between mathematics and everyday experiences. However, the conceptualizations differ substantially: the NRC highlights elements of students' dispositional affective functioning (e.g., beliefs about the worthwhileness and sensibleness of mathematics) as well as conative functioning (e.g., a belief in diligence) that are altogether absent from Fernandez and Cannon's qualification.

Summary Analysis

Several prominent reform-oriented documents suggest that students' dispositions are a crucial consideration for promoting their success in mathematics (e.g., NCTM 1989, 2000; Kilpatrick, Swafford, and Findell 2001). A substantial body of research exists corroborating the recommendations of both NCTM and the NRC, suggesting that students' dispositions are an integral factor influencing how students engage in mathematical activities. However, quite a bit of work remains to explicate the nature of students' dispositions and its impact on students' mathematical activity, because even though the same labels are given to constructs—that is, mathematical dispositions or dispositions toward mathematics—a plethora of interpretations and operationalizations still exists of what constitutes the disposition construct and its components. A first step toward furthering our collective understanding of students' dispositions to mathematics is to develop a shared sense of what "students' dispositions" means as well as a systematic way to map out the vast domain of students' mathematical dispositions and dispositions toward mathematics, that is, their dispositions with respect to mathematics. Perhaps consideration for elements of cognition, affection, and conation can help further our understanding of students' dispositions with respect to mathematics by providing an organizational framework for thinking about dispositional factors in mathematics education.

Organizing elements of the disposition construct identified in the literature, in the context of the cognitive, affective, or conative mode of mental functioning, seems useful. Consider one feature of students' mathematical dispositions, discerning mathematically acceptable explanations, discussed earlier. How someone might discern the mathematical acceptability of explanations could vary (e.g., employing heuristics to govern all comparisons or approaching each case as a unique comparison); the process of discerning mathematically acceptable explanations is arguably a mental function that generates an awareness in the individual of new knowledge. For example, one can think of discerning mathematically acceptable explanations as a dispositional cognitive function.

Also, consider the prevalent references to students' attitudes toward and beliefs about mathematics as features of students' dispositions with respect to mathematics. Attitudes and beliefs, in the context of the affective domain, are general reactions toward something, the essential quality of an emotion, feeling, mood, or temperament (McLeod 1992); that is, they represent elements of dispositional affective functioning. Last, consider the notion of persistence or diligence, both of which various researchers highlighted as components of students' dispositions with respect to mathematics. Conative mental functions are described as the aspect of mental process that is directed toward action or change; impulse, volition, desire, striving, and so forth, all emphasize the conative aspect. Persistence and diligence both imply a directed, purposeful action, that is, dispositional conative functions. Considering features of students' dispositions with respect to mathematics as elements of the cognitive, affective, or conative mode of mental functioning may afford one a more systematic way to organize and conceptualize students' dispositions with respect to mathematics as dispositional cognitive, affective, or conative functions.

Discussion

This review synthesized research on conceptualizations and operationalizations of students' dispositions with respect to mathematics, drawing from educational psychology and math education. I derived a conceptual framework for students' dispositions with respect to mathematics on the basis of categories of mental functioning, such as cognition, affection, and conation. This review's categorization scheme presents a common language and framework for discussing and organizing conceptualizations and operationalizations of the disposition construct, as well as a lens for exploring the nature of students' dispositions in relation to mathematics. At times, mathematics educators and researchers seemed to lack consistency and coherence in discussing and using the disposition construct. Two umbrella terms were prevalent in the literature: *mathematical dispositions* and *dispositions toward mathematics*; however, researchers often conceptualized these terms and their components in substantially different ways. Some attended only to dispositional affective and conative functions, whereas others included cognitive dispositional functions. This review's conceptual framework affords both researchers and educators a more coherent and cohesive means for discussing and mapping the vast domain of students' dispositions with respect to mathematics in light of dispositional cognitive, affective, and conative functions.

Research on the nature of students' dispositions with respect to mathematics offers compelling evidence that students' dispositional cognitive, affective, and conative functioning can influence their learning in mathematics. An absence of certain dispositional cognitive functions can limit students' opportunities to extend their mathematical knowledge and understanding. Students' dispositional affective functioning can negatively influence their learning in mathematics by limiting access to opportunities to learn or by affecting how students take advantage of opportunities to learn. Students' dispositional conative functioning can affect the level of interaction with opportunities to learn, such as challenging, nonroutine problems.

Where might we go from here? Research shows that isolated components of students' dispositions can influence their engagement in mathematical tasks; however, precisely how or to what extent is not entirely clear. This review's framework allows for a more comprehensive approach to addressing the question of how students' dispositions are related to their learning of mathematics. We now can look at the relative contribution of each dimension of students' dispositions with respect to mathematics to their learning of mathematics. For example, we could examine whether cognitive, affective, or conative dispositional functions contribute more effectively to students' development of mathematical knowledge.

The framework also affords researchers and educators a lens with which to examine the development of students' dispositional cognitive, affective, and conative functioning, as well as to identify potential gaps in the literature. Research suggests that several factors contribute to the development of students' dispositional cognitive and affective functioning. Several study findings indicate that teachers' attitudes toward and beliefs about mathematics, as well as their classroom practices, play a significant role in shaping students' dispositional affective and cognitive functioning. However, factors contributing to the development of students' dispositional conative functioning are not as well documented.

REFERENCES

Alpert, Richard, and Ralph Haber. "Anxiety in Academic Achievement Situations." *Journal of Abnormal and Social Psychology* 61 (September 1960): 207–15.

Berliner, David, and Bruce Biddle. *The Manufactured Crisis: Myth Fraud and the Attack on America's Public Schools*. New York: Addison-Wesley, 1995.

Boaler, Jo. "Exploring the Nature of Mathematical Activity: Using Theory Research and 'Working Hypotheses' to Broaden Conceptions of Mathematics Knowing." *Educational Studies in Mathematics* 51 (July 2002): 3–21.

Dweck, Carol. "Motivational Processes Affecting Learning." *American Psychologist* 41 (October 1986): 1040–48.

English, Horace, and Ava English. *A Comprehensive Dictionary of Psychological and Psychoanalytic Terms*. New York: Longman's Green, 1958.

Fernandez, Clea, and Joanna Cannon. "What Japanese and U.S. Teachers Think About When Constructing Mathematics Lessons: A Preliminary Investigation." *Elementary School Journal* 105 (May 2005): 481–96.

Hiebert, James. *Conceptual and Procedural Knowledge: The Case of Mathematics*. Mahwah, N.J.: Lawrence Erlbaum, 1986.

Kilpatrick, Jeremy, Jane Swafford, and Bradford Findell, eds. *Adding It Up: Helping Children Learn Mathematics*. Washington, D.C.: National Academies Press, 2001.

McClain, Kay, and Paul Cobb. "Supporting Students' Ability to Reason about Data." *Educational Studies in Mathematics* 45 (March 2001): 103–29.

McIntosh, Margaret. "Formative Assessment in Mathematics." *Clearing House* 71 (November–December 1997): 92–96.

McLeod, Douglas. "Research on Affect in Mathematics Education: A Reconceptualization." In *Handbook of Research on Mathematics Teaching and Learning*, edited by Douglas Grouws, pp. 575–96. Reston, Va.: National Council of Teachers of Mathematics, 1992.

Moldavan, Carla, and Lelia Mullis. *Fostering Disposition toward Mathematics*, 1998. ERIC Document Reproduction no. ED421352.

National Council of Teachers of Mathematics (NCTM). *Curriculum and Evaluation Standards for School Mathematics*. Reston, Va.: NCTM, 1989.

———. *Principles and Standards for School Mathematics*. Reston, Va.: NCTM, 2000.

Noss, Richard, Lulu Healy, and Celia Hoyles. "The Construction of Mathematical Meanings: Connecting the Visual with the Symbolic." *Educational Studies in Mathematics* 33 (July 1997): 203–33.

Royster, David, Kim M. Harris, and Nancy Schoeps. "Dispositions of College Mathematics Students." *International Journal of Mathematical Education in Science and Technology* 30 (May 1999): 317–33.

Schoenfeld, Alan. "Explorations of Students' Mathematical Beliefs and Behavior." *Journal for Research in Mathematics Education* 20 (July 1989): 338–55.

Skemp, Richard. "Relational Understanding and Instrumental Understanding." *Mathematics Teaching* 77 (December 1976): 65–71.

Snow, Richard E., Lyn Corno, and D. Jackson III. "Individual Differences in Affective and Conative Functions." In *Handbook of Educational Psychology*, edited by David Berliner and Robert Calfee, pp. 243–310. New York: Simon & Schuster Macmillan, 1996.

Part II
Cultural and Societal Issues

PART II of this yearbook is titled "Cultural and Societal Issues" and features four articles. This section explores different ways that cultural and societal issues affect students' motivation to learn mathematics. This part of the yearbook also examines the culture of the mathematics classroom in light of how it affects students' attitudes and beliefs about mathematics and their future careers. Also implicit in this section is the importance of "an equitable distribution of material and human resources; intellectually challenging curricula; educational experiences that build on students' cultures, languages, home experiences, and identities; and pedagogies that prepare students to engage in critical thought and democratic participation in society" (Lipman 2004, p. 3).

Furthermore, throughout this section, authors use specific cases of mathematics classrooms, cultural groups, and students to explicate the role of culture regarding students' motivation to learn mathematics. Through these particular circumstances, we hope that readers will think about their own classrooms and the notion that how mathematics is taught has lasting effects on students' beliefs and attitudes toward mathematics, as well as their motivation to do mathematics and pursue mathematics-related fields.

"Mathematical Dream Makers: How Two Different Math Departments Brought About Equity and High Achievement" compares how mathematics classroom cultures that encourage students to achieve at high levels and see mathematics as connected to their world in meaningful ways differ from classrooms in which students experience mathematics only as rules and rote memorization. Through case studies of particular classrooms and students, the author shows how highly motivational and meaningful teaching of mathematics affects students and that what students experience in high school can have lasting effects on their career choices and what they continue to believe about mathematics.

"What Motivates Students to Take Advanced Mathematics: A Rural Perspective" presents the voices of rural high school students who are motivated to enroll in advanced high school mathematics courses for reasons both expected and unexpected. Through student case studies, the author discusses different types of goals that students may have that stimulate their desire to study advanced mathematics, as well as the implications for teachers. Students with learning goals are motivated to study advanced mathematics because they can improve their skills and knowledge, and studying mathematics has inherent worth for them. Students with performance goals are motivated to do better than others in their math class (i.e., get good grades), in college, or in future career opportunities. Overall, the author wants teachers to see that students' decisions to study advanced mathematics are complex and can be influenced by how mathematics was taught in their preceding courses.

"Factors That Motivate Aboriginal Students to Improve Their Achievement in School Mathematics" focuses on the importance of establishing a reciprocal relationship between schools and students' homes/communities to promote meaningful mathematics learning. The author urges parents, community leaders, teachers, and administrators to work together to ensure that meaningful mathematics learning is taking place. The author also states that mathematics learning should occur in real-world places that use mathematics, not just in the classroom. For evidence of the kind of mathematics learning that it advocates, the chapter uses voices from the Aboriginal culture.

Finally, in "What Motivates Mathematically Talented Young Women?" the authors share their work from a study that investigated attribution and self-efficacy of female high school students through journals and interviews after they participated in a mathematics camp. The authors found that motivation for and disposition toward mathematics as well as environmental factors influence whether mathematically talented young women pursue mathematics. The authors discussed the importance of recognizing early the talents of mathematically gifted young women and placing the girls early on and throughout their schooling in situations that can strengthen their talents.

While reading the papers in this section, consider some of these questions:

- What pedagogical strategies lead to all students' understanding and pursuing more mathematics?
- What factors motivate students from different cultures to do mathematics?
- What professional development do teachers need to help transform their teaching from more traditional pedagogy to pedagogy that is more inclusive, respectful of students' cultures, and challenging to all students?
- How can we help teachers develop reciprocal relationships with parents and communities that allow students to value their home and school cultures in the context of teaching and learning mathematics?
- What mechanisms need to be in place to foster positive mathematics dispositions in all students?
- What lessons can teachers learn from this section's case studies that can apply to motivating all students to learn mathematics?

Marilyn E. Strutchens

REFERENCE

Lipman, Pauline. *Regionalization of Urban Education: The Political Economy and Racial Politics of Chicago-Metro Region Schools*. Paper presented at the annual meeting of the American Educational Research Association, San Diego, April 2004.

Chapter 6

Mathematical Dream Makers: How Two Different Math Departments Brought About Equity and High Achievement

Jo Boaler

One of the most important questions for our time concerns how we may teach students mathematics to counter the "savage inequalities" (Kozol 1992) that prevail in society. All teachers are aware that students come to them with different understandings, different motivations to learn, and different support structures in their homes and lives. Unfortunately, many teachers believe that these differences justify the unequal journeys that students experience in education and the vastly different records of achievement that result, disturbingly predictable by race; social class; and, at some levels, gender (Gutiérrez 2002). This chapter will describe how two math departments, one in the United States and one in the United Kingdom, fought inequalities through their teaching and the incredible results that they achieved. In both cases the equitable outcomes came about through the messages that teachers communicated to students—that everyone can be smart and achieve at the highest levels—with support from innovative teaching approaches designed to encourage all students to high levels.

The Project-Based Approach

This section describes my observations and experiences at two schools with different approaches to teaching mathematics.

Phoenix Park School

The day that I walked into Phoenix Park School, in a working-class area of England, I didn't know what to expect. I had invited the mathematics department at the school to be part of my research project into the effectiveness of different mathematics approaches. I knew that the department used a "project-based approach," but I did not know much more than that. For my research, I monitored students between thirteen and sixteen years of age. To find out about students' experiences, I watched more than 100 hours of lessons in each school, surveyed students every year, interviewed students and teachers regularly, and administered various assessments to find out what the students were learning. I contrasted Phoenix Park with another school, Amber Hill, that used a more typical, traditional approach. I chose the two schools not only because of their different approaches but also because the student intakes were demographically similar; the teachers were well qualified; and the students had followed identical mathematics approaches up to the age of thirteen, when my research began. At that time, students at the two schools scored at the same levels on national mathematics tests. Then their mathematical pathways diverged, with one group attending a school using a traditional teaching approach and the other students attending a school that taught mathematics differently. The next three years, during which I conducted research on the students' experiences, proved extremely interesting.

The classrooms at Phoenix Park looked chaotic. The mathematics approach was project based, which meant a lot less order and control than in traditional approaches. Instead of teaching procedures that students would practice, the teachers gave the students projects to work on that needed mathematical methods. From the beginning of year eight (when students started at the school) to three-quarters of the way through year ten, the students worked on open-ended projects in every lesson. The students didn't learn separate areas of mathematics, such as algebra or geometry, because UK schools do not separate mathematics in that way; instead, they learned "maths," the whole subject, every year. The students learned in mixed-ability groups, rather than in tracks or sets, as we call them in the UK, and projects usually lasted for about three weeks.

At the start of the different projects, the teachers would introduce students to a problem or a theme that the students explored, using their own ideas and the mathematical methods that they were learning. The projects usually had an open structure so that students could take the work in directions of interest to them. For example, a project called "volume 216" simply told the students that the volume of an object was 216, and they had to go away and think about what the object could be, what dimensions it would have, and what it would look like. Sometimes teachers taught mathematical content that could be useful to students before the start of a new project. More typically, the teachers would introduce methods to individuals or small groups when students encountered a need for those methods in a particular project. Sometimes the different projects varied in difficulty, and the teachers guided students toward projects suited to students' strengths. Importantly, the teachers did not predetermine what students could do, as happens in ability groups; instead, teachers communicated that all students could attempt work of any challenge level, and the teachers constantly encouraged students to the highest levels.

During one of my classroom visits, students were working on a project called "36 fences." The assignment asked students to find the maximum area of a space that a farmer could make with 36 equal-sized fences. Some students set about investigating different-sized rectangles, whereas others investigated the size of different shapes. In one group, two boys had found that the largest area came from a thirty-six-sided shape, but they did not know how to find the exact area of the shape. Then the teacher taught the boys to use trigonometry, and they excitedly set about using their new math tools to find the area of each triangle in the shape and then the area of the whole shape.

At Phoenix Park, the teachers taught mathematical methods to help students solve problems. Students learned about statistics and probability, for example, as they worked on a set of activities called "Interpreting the World." During that project they interpreted data on college attendance, pregnancies, sports results, and other issues of interest to them. Students learned about algebra as they investigated different patterns and represented them symbolically; they learned about trigonometry in the "36 fences" projects and by investigating the shadows of objects. The teachers carefully chose different projects to interest the students and to give opportunities for learning important mathematical concepts and methods. Some projects had an applied nature, requiring that students engage with real-world situations; other activities started with a context, such as 36 fences, but led to abstract investigations. As students worked, they learned new methods, chose between different methods, and adapted and applied methods. Not surprisingly, the Phoenix Park students came to view mathematical methods as flexible problem-solving tools, and they learned to adapt and apply methods to fit different situations. A second-year Phoenix Park student, Lindsey, described the math approach: "Well, if you find a rule or a method, you try and adapt it to other things. When we found this rule that worked with the circles, we started to work out the percentages and then adapted it, so we just took it further and took different steps and tried to adapt it to new situations."

Amber Hill School

At Amber Hill School, the teachers used the traditional approach that is commonplace in UK and U.S. schools. The teachers began lessons by lecturing from the board, introducing students to mathematical methods; students would then work through exercises in their books, practicing the methods. The students were placed into one of eight sets (ability groups) for math lessons.

Classrooms at Amber Hill were peaceful and quiet, and students worked quietly, on task, for almost all their lessons. During the three years that I monitored the students as they progressed through school, I learned that the students worked hard but that most of them disliked mathematics. They came to believe that math was a subject that involved only memorizing rules and procedures, as Simon described to me: "In maths, there's a certain formula to get to, say, from *a* to *b*, and there's no other way to get to it; or maybe there is, but you've got to remember the formula—you've got to remember it." More worryingly, the students at Amber Hill became so convinced of the need to memorize the methods they learned that many of them saw no place for thought, as Louise, a student in the highest group, told me: "In maths you have to remember; in other subjects, you can think about it."

Amber Hill's approach contrasted sharply with Phoenix Park's: the Amber Hill students spent more time on task, but they thought that math was a set of rules to memorize, and few students developed the levels of interest that the Phoenix Park students did. In lessons, the Amber Hill students were often successful, getting many correct answers in their exercises, not necessarily by understanding the mathematical ideas but by following cues. For example, the biggest cue telling students how to answer a question correctly was the method that the teacher had just shown on the board. The students knew that if they used the method they had just seen, they were probably all right. They also knew that when they moved from exercise A to exercise B, they should do something slightly more complicated. Another cue they followed was their belief that they should use all the lines in a diagram and all the numbers in a question; if they didn't use them all, they would think that they were doing something wrong. The mathematics classrooms at Amber Hill always followed the same routines, and students came to believe that math *was* the following of routines. Unfortunately, when the students faced their national exams, the cues that they had learned to rely on in class were not there. Gary explained the difficulty that he faced in the national exam: "It's different, and, like, the way it's there, like—not the same. It doesn't, like, tell you it, the story, the question; it's not the same as in the books, the way the teacher works it out." Gary seemed to suggest, as I had seen in my observations, that the story or the question in their books often gave away what they had to do, but the exam questions didn't. Trevor also talked about cues when he explained why his exam grade hadn't been good: "You can get a trigger, when she says, like, 'simultaneous equations' and 'graphs' or 'graphically.' When they say, like, and you know, it pushes that trigger, tells you what to do." I asked him, "What happens in the exam when you haven't got that?" He gave a clear answer: "You panic."

Measures of Achievement

In England, all students take the same national examination in mathematics at age sixteen. The examination is a three-hour, traditional test made up of short mathematics questions of the type that Amber Hill students practiced in every lesson. At Phoenix Park, the teachers stopped the project work a few weeks before the examination and focused on teaching any standard methods that students may not have met. They spent more time lecturing from the board and gave students questions to practice, so for a few months the classrooms looked similar to those at Amber Hill.

Many people expected the Amber Hill students to do well on the examinations, given the school's examination-oriented approach, but the Phoenix Park students attained significantly higher examination grades. The Phoenix Park students didn't do better only on the examinations. For part of my research I investigated the usefulness of the approach to students' lives. Interviewing the students at both schools about their use of mathematics outside school, I found stark differences. All forty Amber Hill students whom I interviewed said that they would never use their school-learned methods outside school. As Richard told me, "Well, when I'm out of school, the maths from here is nothing to do with it, to tell you the truth. . . . Most of the things we've learned in school we would never use anywhere."

At Phoenix Park, the students were confident that they would use the methods that they learned in school, and they could give me examples of using school-learned mathematics in their jobs and lives. Indeed, many of the students' descriptions suggested that they had learned mathematics in a way that transcended the boundaries (Lave 1996) that generally exist between the classroom and real situations.

Teaching for Equity

The results from Phoenix Park were impressive in many respects, particularly since the school was based in one of the most deprived areas of the country and most of the students lived on a housing estate (similar to housing projects in the United States) where police would not venture at night. The students started the school at significantly lower levels than the national average, but they left the school at significantly higher levels. And whereas no gender differences in performance existed at Phoenix Park, significantly more boys achieved the highest grades at Amber Hill. Even more remarkable, perhaps, were my results from measuring social class. I had determined the occupations of all the parents in the two schools and ranked them on a scale of social class. This analysis showed that at the beginning of school, achievement in both Amber Hill and Phoenix Park correlated with social class, with students from lower social class groups achieving at lower levels than those from higher groups, as is typical in schools (Gorard, See, and Smith 2008). But when the students left Phoenix Park, the social class differences in achievement had disappeared. At Amber Hill, the students' achievement on the national exam correlated with the students' social class, with students from lower social groups achieving at lower levels. Phoenix Park had no such correlations; instead, students' achievement related to the amount of effort that they put into lessons.

A major source of the equitable and high achievement at Phoenix Park was the mixed-ability grouping that teachers used, encouraging all students to high levels. At Amber Hill, the students had been put into one of eight sets, ordered by achievement, and teachers taught to the average of the groups. Many students from middle and lower sets in Amber Hill told me that they gave up when they realized that they could not aim for the highest grades. At Phoenix Park, the teachers gave the students work that was open enough for them to take their projects in different directions and different levels. Students worked at an appropriate level for their understanding of a particular area, and this approach encouraged all students to achieve at the highest levels. At Phoenix Park, the teachers believed that all students could achieve highly and constantly communicated this message to students. Some students started school at a low level and would have been put into one of the lowest sets at Amber Hill, resulting in low achievement three years later. But at Phoenix Park they flourished and achieved the highest possible exam grades.

During my study I investigated the sources of both social class and gender equality at Phoenix Park. The source of gender equity, for interest and achievement, came about because teachers at Phoenix Park gave students opportunities to develop a connected understanding of mathematics. As a professor of math education, I have observed thousands of math classrooms, and I have often seen girls and boys reacting differently to math. Math teaching approaches that are equally suitable for girls and boys exist, such as Phoenix Park's approach, but the traditional version of math prevalent in most classrooms is much more suitable for boys than girls—which is why many girls choose not to go forward in math, despite their high achievement (Boaler 2008). Leading-edge research on the different ways that female and male brains process problems (Boaler 2008) has revealed that girls, in particular, need to see the connections between math and the world, between numbers and visual images, and between different areas of math. Many boys appreciate learning a connected version of math, too—and for girls and boys to understand the connections that make up math is good—but most girls and women need the connected version of the subject more urgently than boys do and underachieve when it is denied to them. When boys learn a disconnected, abstract version of math, they tend to "just do it," even though it is neither very interesting for them nor a good way of learning, whereas women and girls will strive for a connected understanding and reject math if such understanding does not become available to them. Comments from girls at Amber Hill often reflected this idea. For example, Jane said, "He'll write it on the board and you end up thinking, 'Well, how come it's this and this? How did you get that answer? Why did you do that?'" Another girl, Gill, said, "It's like, you have to work it out and you get the right answers, but you don't know what you did; you don't know how you got them, you know?"

Amber Hill's abstract, disconnected approach caused girls to underachieve and turn away from the subject. At Phoenix Park, students had opportunities to discuss different methods and to gain a connected understanding of mathematics; this approach helped both girls and boys, and it resulted in equitable achievement.

Mathematics for Life?

When I finished my study of the teaching approaches at Amber Hill and Phoenix Park, I had thought that it was the last I would see of the students from the two schools. But some eight years later I was presenting the results of the study in the United States. A businessman from Texas urged me to find the students who had attended the two schools and investigate how the different teaching approaches had affected their lives. So I returned to England and contacted the ex-students, asking them to complete a survey and then follow-up interviews. The survey asked the young people, who were then around twenty-four years of age, what jobs they were doing. I then classified all the jobs and ranked them by social class. Doing so showed something interesting: When the students had been in school, their social class levels (determined from parents' jobs) had been equal. Eight years after my study, the Phoenix Park young adults were working in jobs that put them at significantly higher social class levels than the Amber Hill adults, even though the school achievement range of those who had replied to the surveys from the two schools had been equal. Sixty-five percent of the Phoenix Park adults had increased their social class categorization, compared with 23 percent of Amber Hill adults. Fifty-two percent of the Amber Hill adults were now in a lower social class category than their parents, compared with only 15 percent of the Phoenix Park adults. Phoenix Park showed a distinct upward trend in social class among the children, whereas Amber Hill had no such trend, even though Phoenix Park was in a less prosperous area.

In addition to the survey, I conducted follow-up interviews with the young adults. I contacted a representative group from each school, choosing young adults with comparable examination grades from each school. In interviews, the Phoenix Park adults communicated a positive approach to work and life, describing how they used the problem-solving approach that they had learned in their mathematics classrooms to solve problems and make sense of mathematical situations in their lives. When I asked Paul, a senior regional hotel manager, whether he found the mathematics that he had learned in school useful, he said that he did: "I suppose there was a lot of things I can relate back to maths in school. You know, it's about having a sort of concept, isn't it, of space and numbers and how you can relate that back. And then, okay, if you've got an idea about something and how you would then use maths to work that out. . . . I suppose maths is about problem solving for me. It's about numbers, it's about problem solving, it's about being logical."

Whereas the Phoenix Park young adults talked of math as a problem-solving tool and were generally positive about their school's approach, the Amber Hill students could not understand why their school's mathematics approach had prepared them so poorly for the demands of the workplace. Bridget spoke sadly when she said, "It was never related to real life, I don't feel. I don't feel it was. And I think it would have been a lot better if I could have seen what I could use this stuff for, and just basically . . . because then it helps you to know why. You learn why that is that, and why it ends up at that. And I think definitely relating it to real life is important." Marcos was also puzzled why the school's math approach had seemed so removed from the students' lives and work: "It was something where you had to just remember in which order you did things, and that's it. It had no significance to me past that point at all—which is a shame. It was very abstract. And it was kind of almost purely theoretical. As with most things that are purely theoretical, without having some kind of association with anything tangible, you kind of forget it all."

Notably, in the interviews with the Amber Hill ex-students, the aspect of school that they were most keen to talk about was how ability grouping had shaped their entire educational experience. Amber Hill students from set 2 downward talked not only about how the grouping had constrained their achievement but also how it had set them up for low achievement in life. In interviews, the students from Phoenix Park talked about how the school had excelled at finding and promoting the potential of different students and that teachers had regarded everyone as a high achiever. The young adults communicated a positive approach to work and life, describing how they used the problem-solving approaches that they had learned in school to get on in life. The young adults who had attended Amber Hill and had been put into sets told me that their ambitions were "broken" at school and their expectations lowered. One young man spoke eloquently about the ability grouping experience: "You're putting this psychological prison around them . . . it's kind of . . . people don't know what they can do, or where the boundaries are, unless they're told at that kind of age."

Ability grouping affects students' lives profoundly, both in and beyond school. Researchers in England (Dixon 2002) have revealed that 88 percent of children placed into ability groups at age four remain in the same groupings until they leave school. This statistic is one of the most chilling that I have ever read. The fact that children's futures are decided for them by the time they are placed into groups, at an early age, derides the work of schools and contravenes basic knowledge about child development and learning. Children develop at different rates, and they reveal different interests, strengths, and dispositions at various stages of their development. One of the most important goals of schools is to furnish stimulating environments for all children: environments that can pique and nurture children's interest, with teachers who are ready to recognize, cultivate, and develop the potential that children show at different times and areas. The fact that the Phoenix Park teachers did not prejudge the students' potential and that they used multileveled mathematics materials that all students could take to their own highest level meant that students could all maintain high aspirations. Such aspirations seemed to encourage them not only in school but also in life.

The Communicative Approach

After I moved to the United States, I conducted a follow-up study of California high schools using different math approaches. I could not collect social class data, but the data I collected on the relationship between ethnicity and achievement in one school produced stunning conclusions.

Railside High School is an urban California high school. The sound of speeding trains, passing not far from the classrooms, often interrupts lessons. As with many urban schools, the buildings look as though they need some repair. But Railside is not like other urban schools in all respects. Calculus classes in urban schools often have poor enrollment or are nonexistent, but at Railside, eager and successful students pack these classes. Visitors I brought to the math classrooms were amazed when they saw all the students hard at work—engaged and excited about math. I first visited Railside because I had learned that the teachers collaborated and shared teaching ideas, and I was interested to see their lessons. I saw enough in that visit to invite the school to be part of a new Stanford University research project investigating the effectiveness of different mathematics approaches. Some four years later, after we had followed

700 students through three high schools—observing, interviewing, and assessing students—we knew that Railside's approach was both highly successful and highly unusual.

The mathematics teachers at Railside used to teach using traditional methods, but the teachers were unhappy with the high failure rates among students and the students' lack of interest in math. So they worked together to design a new approach. Teachers met together over several summers to devise a new algebra curriculum and later to all the math courses available. They also detracked classes and made algebra the first course that all students would take on entering high school. In most algebra classes, students work through questions designed to give practice on mathematical techniques such as factoring polynomials or solving inequalities. At Railside, the students learned the same methods, but the curriculum was organized around bigger mathematical ideas, with unifying themes such as "What is a linear function?" Also, the students' mathematical work was highly communicative: the students learned about the different ways to communicate mathematics, such as through words, diagrams, tables, symbols, objects, and graphs. Teachers would often ask students to explain work to each other, moving among different representations and communicative forms. This was why, when we interviewed students and asked them what they thought math was, they did not tell us that it was a set of rules, as most students do; instead, they told us that math was a form of communication, or a language, as Jose explained: "Math is, like, kind of a language, because it has got a whole bunch of different meanings to it, and I think it is communicating. When you know the solution to a problem, I mean, that is kind of like communicating with your friends."

The tasks at Railside were designed to fit within the separate subject areas of algebra and geometry, in line with the tradition of content separation in U.S. high schools. The problems that the students worked on were open enough to allow thinking in different ways, they often required that students indicate their thinking with different mathematical representations, and they emphasized the connections between algebra and geometry.

The Railside classrooms were all organized in mixed-ability groups, and teachers used a pedagogical approach called "complex instruction" to make the group work equal and to encourage students to help each other as they worked (Boaler 2008). The Railside teachers paid a great deal of attention to how the groups worked, and they taught students to respect the contributions of other students, regardless of social status or prior attainment. One unfortunate but common side effect of some classroom approaches is that students develop beliefs about the inferiority or superiority of different students. In the other, traditionally taught, classes that we studied, students talked about other students as smart and dumb, quick and slow. At Railside, the students did not talk in these ways. This did not mean that they thought all students were the same—they did not—but they came to appreciate the diversity of classes and the different attributes that different students offered, as Zane described: "Everybody in there is at a different level. But what makes the class good is that everybody's at different levels, so everybody's constantly teaching each other and helping each other out."

For part of our research project, we compared the learning of the Railside students to that of a similar-sized group of students in two other high schools. These other students were learning mathematics through a more typical, traditional approach. In the traditional classes, students sat in rows at individual desks, they did not discuss mathematics, they did not represent algebraic relationships in different ways, and they generally did not work on applied or visual problems. Instead, the students watched the teacher demonstrate procedures at the start of lessons and then worked through textbooks filled with short procedural questions. The two traditional schools were more suburban, and students started the schools with higher mathematics achievement levels than did the students at Railside. By the end of the first year of our research study, the Railside students were achieving at the same levels as the students in the more suburban schools on tests of algebra; by the end of the second year, the Railside students were outperforming the other students on tests of algebra and geometry. In addition, when the students started at Railside, students of different ethnic groups scored at significantly different levels. But by the time the students left Railside, all the differences between ethic groups had significantly diminished and in some cases disappeared. Also, girls and boys achieved at the same levels.

In addition to the high and equitable achievement at Railside, the students were also extremely positive about mathematics. In various surveys during the four years of the study, the students at Railside were always significantly more positive and more interested in mathematics than the students from the other classes. By their senior year, a staggering 41 percent of the Railside seniors were in advanced classes of precalculus and calculus, compared with only 23 percent of students from the traditional classes. Further, when we interviewed 105 students at the end of the study, mainly seniors, about their future plans, almost all students from the traditional classes said that they had decided not to pursue mathematics as a subject—even when they had been successful. Only 5 percent of students from the traditional classes planned a future in mathematics, compared with 39 percent of Railside students.

Many reasons account for the success of the Railside students. Importantly, the students had opportunities to work on interesting problems that required them to think, and not just reproduce methods, and they had to discuss mathematics with each other, increasing their interest and enjoyment. Also, the teachers at Railside were careful in identifying and talking to students about all the ways in which they were "smart." The teachers knew that a belief of not being smart enough often severely hampered students—and adults—in their mathematical work. They also knew that all students could contribute a great deal to mathematics, and so they took it upon themselves to identify and encourage all the students' different strengths, in the same way that the Phoenix Park teachers had. This approach paid dividends, and the motivated and eager students who believed in themselves and who knew that they could succeed in mathematics would have impressed any visitor to the school.

Conclusion

The two teaching approaches that I have reviewed were the subject of comprehensive and longitudinal research studies, and although I conducted them in different countries, the findings were consistent. Despite the prevalence of traditional teaching methods, almost all research studies on mathematics learning point to the same conclusion: students need to be actively involved in their learning, and they need to be engaged in a broad form of mathematics—using and applying methods, representing and communicating ideas—to be interested and engaged.

The different studies also showed that a major source of social and cultural equity came about through grouping students, and when students were grouped together, irrespective of prior attainment, and all students were encouraged to high levels, traditional sources of inequality dissipated and achievement significantly increased. When teachers encouraged all students to high levels, through using open mathematics materials and through communicating the message that all students can achieve at high levels through hard work, traditional barriers broke down. Many teachers believe that children's school achievement and progress will inevitably reflect societal inequalities. The Phoenix Park and Railside teachers did not hold such beliefs, and in their work to promote mathematical equity they gave students opportunities that changed their lives in many different and important ways. All teachers can play such roles, even though changing teaching methods takes a lot of courage, especially without support. Ladson-Billings (1994) described teachers as "dream keepers," but for the students whom they believe in and promote, teachers who enact equitable teaching approaches are nothing short of mathematical dream makers.

REFERENCES

Boaler, Jo. *What's Math Got to Do with It? Helping Children Learn to Love Their Least Favorite Subject—and Why It's Important for America.* New York: Penguin, 2008.

Dixon, Annabelle. "Editorial." *FORUM* 44, no. 1 (2002): 1.

Gorard, S., B. See, and E. Smith. "The Impact of SES on Participation and Attainment in Science: Strand 1—a Review of Available Data." Paper presented at the Royal Society Seminar, Socioeconomic Status and School Science, December 2008.

Gutiérrez, Rochelle. "Enabling the Practice of Mathematics Teachers in Context: Toward a New Equity Research Agenda." *Mathematical Thinking and Learning* 4, no. 2–3 (2002): 145–87.

Kozol, Jonathan. *Savage Inequalities: Children in America's Schools*. New York: Harper Perennial, 1992.

Ladson-Billings, Gloria. *The Dreamkeepers: Successful Teachers of African American Children.* San Francisco: Jossey-Bass, 1994.

Lave, Jean. "Teaching, as Learning, in Practice." *Mind, Culture, and Activity* 3 (Summer 1996): 149–64.

Chapter 7

What Motivates Students to Take Advanced Mathematics: A Rural Perspective

Rick Anderson

WHAT MOTIVATES high school students to enroll in advanced mathematics courses? Is it because they enjoy mathematics? Or are they preparing for future educational or occupational opportunities? According to the National Council of Teachers of Mathematics (NCTM) *Principles and Standards for School Mathematics*, "all students are expected to study mathematics each of the four years that they are enrolled in high school" (NCTM 2000, p. 288), but many students enroll in only the two or three years of mathematics that schools require for graduation. Mathematics teachers have an ongoing concern with students' participation in mathematics courses; many have argued that mathematical knowledge is necessary for *all* individuals since economic, social, and political opportunities depend on it (Moses and Cobb 2001). So what motivates students to enroll in advanced high school mathematics courses beyond the minimum requirements for graduation? What may be specific to students in rural high schools?

Students attending school in small towns or rural areas make up more than 20 percent of public schools (Johnson and Strange 2007). These students have different experiences from those in nonrural areas since rural schools are typically smaller (Beeson and Strange 2003), less likely to offer advanced coursework (Greenberg and Teixeira 1998), and often farther from colleges or universities (Gibbs 1998). For these reasons, the motivation that rural students have to study advanced high school mathematics may differ from that of students in nonrural areas. In this chapter, we hear from students attending a rural high school who are motivated to enroll in advanced high school mathematics courses for reasons both expected and unexpected. The advanced mathematics courses, such as precalculus and calculus, are beyond those that their school requires for high school graduation.

Motivation and Instruction

Motivations are the "reasons individuals have for behaving in a given manner in a given situation" (Middleton and Spanias 1999, p. 66). Students' motivations for studying mathematics form in the interaction of their personal goals and the instructional practices in the classroom (Turner and Patrick 2004). *Learning goals*—a belief that hard work, understanding the material, and collaboration with others are necessary to succeed in mathematics—motivate some students, whereas *performance goals* motivate other students, who value grades and teacher attention. These latter students define their success as superiority over others. Instructional practices in the classroom both create and maintain students' motivational goals (Middleton and Spanias 1999).

Research often contrasts instruction in mathematics classrooms between *teacher centered* (traditional) on the one hand and *student centered* (standards based) on the other (Boaler 1997; Senk and Thompson 2003). Teacher-centered mathematics instruction consists of teachers demonstrating procedures to the class, after which students replicate and practice those procedures on individual exercises. Tasks often focus on procedures over conceptual understanding, and questions usually have one, often numerical, answer. Teacher-centered instruction usually separates mathematical topics into short, one-day sections, often decontextualized and unrelated to students' lives. Such instruction is commonplace in mathematics classes in the United States (Jacobs et al. 2006; Stigler and Hiebert 1999). In contrast, student-centered mathematics instruction involves developing students' mathematical ideas through problem-solving or inquiry-oriented approaches. Teachers facilitate classroom discussion and collaborative group work, questioning students to push them to develop a deeper understanding of the content as they reason through and justify their solutions. Tasks are open ended and focus on conceptual development that supports procedures and methods. Teachers try to connect various mathematical topics and have students relate them to their own experiences. NCTM (2000) promotes this type of instruction to encourage improved mathematics achievement and understanding among students.

Mathematics instruction in the classroom influences students' beliefs about mathematics and its perceived usefulness, along with students' perceptions of their own abilities to learn mathematics. In a teacher-centered mathematics classroom, students often come to believe that the best way to learn mathematics is through memorization. They believe that a primary goal of mathematics is to determine correct answers. Thus, teacher-centered mathematics classrooms may foster students' performance goals; students in such classrooms perceive mathematics to have little inherent usefulness, are less likely to take an interest in the subject matter, and attempt to compete for high grades as symbolic of their learning. Alternatively, students in student-centered mathematics classrooms are more likely to develop learning goals than performance goals. Students tend to believe that mathematics is useful for both their present and their future, take an interest in the subject matter, and often seek to collaborate with others to develop their understanding (Boaler 1997; Middleton and Spanias 1999).

Students' success over peers sustains motivation primarily through performance goals. The student may perceive that innate ability, not effort, accounts for success. Students attribute their failures to a lack of ability, and motivation diminishes. As a result, student interest may decrease in understanding the material and finding its relevance and meaning. In contrast, students with a learning goal orientation are intrinsically motivated because they understand what they have learned. They do not compete with classmates but rather work collaboratively to better understand the content. Failures are not personal weaknesses but instead a result of a poorly chosen strategy or lack of effort. A continued desire to work hard, understand the content, and collaborate sustains students' motivation (Middleton and Spanias 1999).

"Welcome to Our Neck of the Woods"

We now turn to Cedar Valley High School (CVHS; all names are pseudonyms) and stories of three students enrolled in advanced mathematics courses. We will learn what motivates them to study mathematics in the context of their mathematics classes. The description of the Cedar Valley community, the high school, and the mathematics classes will help you better understand the context of the students' perspectives. The account here is part of a qualitative three-month study (Anderson 2006).

Community

Approximately 3,500 residents live in Cedar Valley, a rural community nestled in the forested mountains of the U.S. Pacific Northwest. Cedar Valley is approximately twenty miles from a small city (population, 35,000) and an hour's drive from the nearest metropolitan area with a population of more than one million. Students described their community as small and friendly, and many indicated that they enjoyed outdoor activities such as camping, hunting, and fishing.

The timber industry shaped the economy of Cedar Valley, but after decades of intense logging eliminated most of the old-growth forested areas, that industry employed fewer people. Some residents continue to work logging in the forests, at the sawmill, or in other timber-related jobs, whereas others commute to work in mills, factories, or stores in nearby communities. Tourist services and seasonal work are also available.

School

The K–12 Cedar Valley School graduates approximately forty to fifty students each year, with a cohort on-time graduation rate of more than 80 percent. More than 90 percent of the students at CVHS are white, with 25 percent of students at CVHS eligible for free or reduced-price lunches. At the time of this study, Ms. Jones, the lone high school mathematics teacher, had ten years of experience teaching at CVHS. She was responsible for teaching most of the high school mathematics classes: algebra 1, geometry, algebra 2, and precalculus. As is more common in rural than nonrural schools (Greenberg and Teixeira 1998), CVHS did not usually offer calculus, although adequate demand caused the school to offer it for the first time during this study. The state required all students to complete two years of mathematics courses to graduate. Approximately 20 percent of the juniors and seniors at CVHS were enrolled in either precalculus or calculus. Most of the others did not enroll in any mathematics classes beyond the two required courses.

Mathematics Classes

The mathematics instruction at CVHS was teacher centered in the context of the preceding definition, and in the words of the math teacher, "We've always stayed pretty traditional. We haven't really changed to the really 'out there' hands-on type of programs." Classroom observations and comments from students corroborated this conclusion. For example, Calvin, a senior, summed up a typical day in a recent math class:

> Just go in, have your work done. First the teacher explains how to do it. Like the Pythagorean theorem, for example, she tells you the steps for it. She shows you the right triangle, the leg, the hypotenuse, that sort of thing. She makes us write up notes so we can check back. And then after that she makes us do a couple [of examples for practice] . . . she shows us the answer. She gives us time to work [on assigned exercises]. Do those and after that she shows us the correct way to do it. If we got it right, then we know. She makes us move on and do an assignment [homework].

Typical of teacher-centered mathematics instruction, most class periods began with a time for students to check answers to homework exercises. Next, as Calvin indicated, the teacher demonstrated a new procedure or method. Students replicated and practiced what the teacher did. Then the teacher gave them an assignment and time to work. The teacher led the activities, and students worked individually on mostly short, decontextualized single-answer exercises. The mathematics classes strongly emphasized "answers": the teacher gave the students answers for their homework assignment, the teacher showed them how to get correct answers to questions that they had difficulty with, and students checked answers to the new homework assignment with other students or with a teacher's answer key.

Enrolling in Advanced Mathematics Classes

With this background, we now return to the question, what motivates students to enroll in advanced high school mathematics courses? In high schools such as Cedar Valley, where graduation requirements do not include four years of mathematics classes, students choose whether to continue to study math each year. What influences their decision? What reasons do they give for enrolling in elective advanced mathematics classes?

A written survey of students enrolled in precalculus or calculus asked students to list three reasons why they had taken more than two years of math in high school. Twenty percent of students indicated that they did so because they liked math, math classes, or the math teacher. Nearly half of the students' reasons for enrolling in the classes suggested that learning goals motivated them (Middleton and Spanias 1999). These students wrote statements such as "I am not good at math so I try to do better," "I want to expand my knowledge," and "I want to challenge myself." The students were willing to try to improve their skills and knowledge and attempt to understand more mathematics.

Many other reasons, however, pointed to students' performance goals. These responses indicated that students were motivated to study advanced mathematics because they could be more competitive in postsecondary education or work. Example responses included "More difficult math classes look better on college transcripts" and "It may help me earn scholarships."

The combination of learning and performance goals that the students indicated is not unexpected, although one might wish that learning goals motivated more students to develop their understanding of mathematics through advanced high school study. Many of these responses mirror those of urban black students: "The students . . . contextualized the importance of mathematics and mathematics knowledge in either socioeconomic terms (i.e., necessary to get a 'good' job) or as a prerequisite to the steps leading to a promising future (i.e., necessary to get into college)" (Martin 2000, p. 168). Like these urban students, more than half of the responses from students at CVHS indicated that future educational and career opportunities influenced their decision to enroll in an advanced high school mathematics class.

Students' Stories

Stories from interviews with students give deeper insight and show a range of motivation for students to continue their study of mathematics through four years of high school. As we will learn, Calvin enrolled in mathematics classes because he liked the classes and wanted to be a high school mathematics teacher. Elizabeth said that she enrolled in the math classes because she believed that they made her more qualified for college. An interest in a career with computers motivated Benjamin to study math throughout high school.

Calvin: Math is "like a jigsaw puzzle waiting to be solved."

Calvin was a student in precalculus. His teacher and classmates considered him a good student. Like most of his classmates, Calvin was the first in his family to study advanced high school mathematics. According to Calvin, his father had dropped out of college and was working in a steel mill; his mother cleaned houses. She had finished high school but had not taken many math classes. "My mom doesn't understand [math]," he said. One of the only people Calvin knew who had studied mathematics beyond high school algebra was Ms. Jones, his high school math teacher.

Calvin said that he enrolled in precalculus because he enjoyed math. "I just like to take it because I'm interested in it and I enjoy taking it. It's just fun, I guess. It's just like a big puzzle; you have to solve it," he said. "It's easy. It's like a jigsaw puzzle waiting to be solved. I like puzzles." Math, for Calvin, was something that "fit together." This statement suggests that learning goals motivated Calvin. He enjoyed math and found it worthwhile to study for its own sake. In the teacher-centered mathematics classroom, he found success replicating the teacher's solution methods and determining correct answers. Yet, despite his positive attitude toward mathematics, he did not exude confidence in his abilities. During independent work time during math class, Calvin often checked his answers with a classmate or with the teacher's answer book. Neither Calvin nor his classmates appeared to have strategies for ascertaining the mathematical validity of their responses or, if they did, chose to call on the external authority of the answer book.

Outside math class, Calvin had experienced some uncertainty and obstacles to success in other areas of his life. He enjoyed sports but could not always make the team, sometimes because of injuries. He was also uncertain of his future career, first considering engineering before deciding on becoming a high school mathematics teacher. Calvin decided to be a math teacher partly because precalculus was "probably" his favorite class. "That's the only thing I understand so far," he said. The teacher-centered math classroom at CVHS was an environment where Calvin could demonstrate his abilities, a place where teachers and students recognized his efforts. Calvin may be like some students that Boaler (2000) described: "[They] are happy stepping into the school mathematics world, either because they find success there or because it offers a form of shelter or recluse from the interactional demands of real life" (p. 393). Calvin accurately reproduced the mathematical procedures (e.g., solving equations, factoring expressions) that his teacher demonstrated and did so more quickly than his classmates, so he viewed his performance in math class as one of his strengths. However, his lack of confidence in knowing whether or when the answers were right diminished the security of "right answers."

Elizabeth: "I figured [math] would look good on transcripts."

Whereas his enjoyment of the mathematics classes motivated Calvin, Elizabeth expressed no real interest in the subject matter or what she was learning. Instead, she was motivated to enroll in advanced high school mathematics classes primarily because she believed that they increased her potential to continue her formal education beyond high school. Elizabeth was enrolled in calculus, her fifth math course in four years. She had enrolled in both precalculus and geometry during her junior year and had more math credits than any other student at CVHS. She had taken the math classes because, she said, "I figured it would look good on transcripts." This statement indicated the main reason why Elizabeth enrolled in so many math classes in high school: the performance goal of doing better than other applicants to the university that she had chosen to attend. She received good grades, and when referring to her calculus class, she was quick to point out, "I'm getting the highest grade in that class." This behavior further illustrates her desire to do better than others in her class.

Elizabeth didn't object to calculus, but her favorite class was English. "I really like to write," she said. She also favored the English class "because I can be creative and basically reinvent myself in my writing." Having an avenue to express herself seemed important to Elizabeth. She felt that the calculus class afforded her no such opportunity. When asked whether calculus was a creative class, Elizabeth conceded that the potential for creativity existed: "If I were to use [calculus] in a creative way. But so far, it's just work and tests." This response reflected the teacher-centered instruction of the mathematics classes that she had taken. Elizabeth's motivation to study English reflects a learning goal orientation. She valued what she could learn through the writing process. In mathematics, however, she found no particular relevance to the knowledge and held a performance goal orientation.

Elizabeth's response to her mathematics education mirrors that of students in similar teacher-centered mathematics classes that Boaler (2000) reported.

Students in these studies indicated displeasure with the monotony and lack of meaning in teacher-centered mathematics classes, leading them to eschew mathematics because "they wanted to pursue subjects that offered opportunities for expression, interpretation, and agency" (Boaler and Greeno 2000, p. 187). Students like Elizabeth can feel constrained and restricted when they cannot express themselves and be creative in math classes. Little motivation exists to study advanced math when students feel that doing so is not worthwhile. None of the knowledge from any math class that she took figured prominently in her future. In fact, she said, "I don't think I'd need calculus to be a journalist."

Elizabeth's story highlights her performance goal orientation. She studied advanced math because "it would look good on transcripts" and allowed her to be better than other college applicants. She seemed to hold no learning goals related to mathematics and seemed to see no use for the mathematical knowledge itself. Elizabeth was right that taking elective advanced mathematics classes is good preparation for college. In fact, a major function of mathematics has been that of gatekeeper to future opportunities (e.g., Moses and Cobb 2001). Yet, it seems an unfulfilling reason for her to spend so much time devoted to an activity that she did not find inherently enjoyable. As teacher-centered mathematics instruction plays out at CVHS and countless other high schools in the United States, it does not encourage and develop the creative potential in students' thinking—even though mathematicians consider mathematics to be a creative discipline.

Since many regard enrolling in advanced mathematics classes as necessary preparation for college, doing so can serve as a prerequisite for rural high school students, like Elizabeth, not only for college but also for life outside their rural community. Elizabeth was vocal in her negative characterization of the students and community. She was eager to move away to an urban area. Perhaps students like Elizabeth choose to enroll in advanced math classes in high school because they lead to college and a life outside the rural community. But what about students who either do not wish to go to college or do not wish to leave their community? Without those goals, what reasons do these students have to enroll in advanced math classes? Students at CVHS seemed to have difficulty answering these questions.

Benjamin: "Everything about computers is doing math."

Benjamin, a senior, was a quiet student who had attended Cedar Valley schools since preschool. He planned to attend college and study computer science. Benjamin was more like Calvin than Elizabeth in his reason for enrolling in precalculus class: "Well, I like math." Then he added, "The teacher's great." Nearly all the students, regardless of whether they enjoyed math, indicated that their math teacher, Ms. Jones, was supportive. When asked what he liked about math, he replied, "Having problems to solve. Just getting a problem and figuring out how to get it done." A desire to work hard to develop his knowledge in mathematics motivated Benjamin. He also wanted to use this knowledge for a career in computer science. Hence, he enrolled in a mathematics class each year of high school since "everything about computers is doing math. . . . I think math is [important] for what I'm going into. If you were a logger, you might need to know how big the tree is. But in computers, you have to know every single math detail about what the program does. Most of it's numbers—all the programs are based mostly on numbers." Benjamin's interest in computers left no question in his mind that he would take the math courses. "If you're in computers, you have to do math. You'd have to know it before you get into it. . . . You'd have to learn it one way or the other." Benjamin saw a strong connection between what he was learning in the math classes and its usefulness for working with computers.

Benjamin was possibly more knowledgeable than his classmates about college and the academic requirements for his career since he had an older brother who had recently enrolled at a university. This connection gave him some insight into what preparing for a career working with computers would require. As a result, Benjamin never doubted that he would take math classes throughout high school. "Well, my brother took [geometry and precalculus] and most of my family's taken it. So it's what you need to go to college."

Unlike Elizabeth, Benjamin enjoyed living in a rural area and wanted to live in a rural area in the future. He said that he was comfortable living in a rural area but recognized that the availability of a job could influence his decision. "I've lived in rural areas most of the time. It depends on where the job is. If I get a job in a city, I'd probably commute to the city and live in a rural area." Benjamin touches on a tension between having a job in a city and a desire to live in a rural area. Others (e.g., Hektner 1995) have noted this tension in studies of rural youth.

Benjamin's story illustrates that high school students may study math because it supports their career goals. For Benjamin, this was working with computers. It also leaves the question, what could someone do for a career after studying mathematics? Calvin considered engineering but opted instead for teaching math. Benjamin could not imagine any career for mathematics majors other than teaching mathematics. His response could reflect the occupations of the residents of Cedar Valley. Most students at CVHS were likely to know only one person who had majored in mathematics in college: their math teacher, Ms. Jones. Such a situation is conceivable in any rural community where logging, farming, ranching, fishing, or mining is the dominant industry. Even in rural areas that have expanded employment opportunities to include other industries, few residents have probably studied mathematics as a university major.

Benjamin realized that pursuing a career working with computers might take him away from his roots in a rural community. Still, he had resolved the dilemma by imagining himself living in a rural area and commuting to work in a city. Can all students see such a possibility? On one hand, might some students who wish to remain in rural areas be less motivated to enroll in advanced high school math courses because they believe that it leads to careers that take them away from the community? On the other hand, could the mathematics learned in high school lead to opportunities where students could contribute to the community in which they reside? Students need to recognize the connection of mathematical knowledge to work and life in rural areas.

Lessons to Be Learned

These stories highlight answers to the question, what motivates students to enroll in advanced high school mathematics courses? We have learned that complex and varied reasons motivate students. Those with learning goals are motivated to study advanced mathematics because they can improve their skills and knowledge. They like math and enjoy solving puzzles and working problems. Studying mathematics has inherent worth for these students. Those with performance goals are motivated to do better than others in their math class (get good grades), in college, or in future career opportunities. These reasons reflect students' perception of themselves, their confidence, and personal goals for life and work. We also see that simple explanations for why students enroll in advanced high school math classes are not possible. Yet, mathematics teachers can play a role in motivating students to continue their mathematics education through four years of high school.

Students motivated by learning goals rather than performance goals tend to do better in mathematics. Also, students in student-centered classes are more likely to develop learning goals "since success is defined as attempts to understand mathematics and explain their thinking to others" (Middleton and Spanias 1999, p. 74). We must question why more students do not have a learning goal orientation toward the study of mathematics. Does learning math for its own sake have value? Teachers should work to create classrooms that reflect student-centered instruction.

Figure 7.1 offers suggestions to encourage problem solving, reasoning, communication, and creativity (Anderson 2007; NCTM 2000). Teachers should work to make mathematics meaningful and relevant to students.

Students like Elizabeth would probably become more interested in mathematics for its own sake in a classroom where students work with mathematical tasks that allow them to develop their own strategies for solving mathematical problems and make sense of mathematical ideas. She would have opportunities to be creative and explain her thinking to others in the class. It may take the focus away from mathematics as something to "get on the transcript" to one where ideas are engaging along with being relevant to her present and future life. But does this type of classroom serve to disadvantage students like Calvin who thrived in the perceived certainty of teacher-centered mathematics instruction? Although he may not be able to check his answers with the teacher or the answer key, he may develop more confidence in his own abilities. In the teacher-centered classroom he seemed to have no way to know the mathematical validity of his

Learning Goals	Performance Goals
• Focus on process and explanations of problem solving rather than emphasize quick responses to single-answer exercises. • Organize mathematics classrooms to allow students to express themselves creatively and communicate their meanings of mathematical concepts to their peers and teacher. • Choose mathematical tasks that allow students to develop strategies for solving problems and meanings for mathematical tools. • Make explicit how mathematics is part of students' daily lives. Help students identify ways to create and use mathematics in their work and play.	• Acknowledge the benefits of advanced mathematics classes and encourage students to take four years of high school mathematics for improved college and career opportunities. • Reward and recognize students who complete advanced mathematics classes. • Commend effort, hard work, and persistence in mathematical problem solving.

Fig. 7.1. Characteristics of learning and performance goals

responses. If part of his learning were to justify, reason, and explain strategies and results, he might well gain more confidence in his own abilities.

Teachers must also be aware of students' performance goals that motivate them to study advanced mathematics. For example, many students may enroll in advanced math classes because of their goals to be the best applicant for work, postsecondary education, or other situations even if mathematics in itself was not rewarding. Knowing this, teachers and school guidance counselors can capitalize on students' existing goals and keep students informed of the educational and occupational opportunities and rewards available to those with a strong high school mathematics background. Such information may motivate some students to continue their study of math beyond the minimum required for graduation.

Finally, mathematics teachers must continue to investigate what motivates students to enroll in advanced mathematics courses. Like the students in this article, each student has her or his own reasons for acting. The following questions may be useful for teachers to consider for their own teaching and students:

- How can enrollment in advanced mathematics classes help students achieve their future postsecondary, career, and personal goals? Are students aware of these benefits?
- How can learning mathematics be meaningful to students in the moment? Do ways exist to make learning relevant to their present interests and not just look to the future?
- How can we learn more about our students' interests and goals and use that knowledge to organize our teaching and classrooms to motivate more students?

When teachers discover what motivates their students, they can support the students to continue their mathematics education, opening possibilities and opportunities for each one and, hopefully, enriching the students' lives as they better understand mathematics.

REFERENCES

Anderson, Rick. "Mathematics, Meaning, and Identity: A Study of the Practice of Mathematics Education in a Rural High School." Ph.D. diss., Portland State University, 2006. *Dissertation Abstracts International* 68-03A: 918.

———. "Being a Mathematics Learner: Four Faces of Identity." *Mathematics Educator* 17 (Summer 2007): 7–14.

Beeson, Elizabeth, and Marty Strange. "Why Rural Matters 2003: The Continuing Need for Every State to Take Action on Rural Education." *Journal of Research in Rural Education* 18 (Spring 2003): 3–16.

Boaler, Jo. *Experiencing School Mathematics: Teaching Styles, Sex, and Setting.* Philadelphia: Open University, 1997.

———. "Mathematics from Another World: Traditional Communities and the Alienation of Learners." *Journal of Mathematical Behavior* 18 (June 2000): 379–97.

Boaler, Jo, and James G. Greeno. "Identity, Agency, and Knowing in Mathematics Worlds." In *Multiple Perspectives on Mathematics Teaching and Learning*, edited by Jo Boaler, pp. 171–200. Westport, Conn.: Ablex, 2000.

Gibbs, Robert M. "College Completion and Return Migration among Rural Youth." In *Rural Education and Training in the New Economy: The Myth of the Rural Skills Gap*, edited by Robert M. Gibbs, Paul L. Swaim, and Ruy Teixeira, pp. 61–80. Ames, Iowa: Iowa State University Press, 1998.

Greenberg, Elizabeth J., and Ruy Teixeira. "Educational Achievement in Rural Schools." In *Rural Education and Training in the New Economy: The Myth of the Rural Skills Gap*, edited by Robert M. Gibbs, Paul L. Swaim, and Ruy Teixeira, pp. 23–39. Ames, Iowa: Iowa State University Press, 1998.

Hektner, Joel M. "When Moving Up Implies Moving Out: Rural Adolescent Conflict in the Transition to Adulthood." *Journal of Research in Rural Education* 11 (Spring 1995): 3–14.

Jacobs, Jennifer K., James Hiebert, Karen Bogard Givvin, Hilary Hollingsworth, Helen Garnier, and Diana Wearne. "Does Eighth-Grade Mathematics Teaching in the United States Align with the NCTM *Standards*? Results from the TIMSS 1995 and 1999 Video Studies." *Journal for Research in Mathematics Education* 37 (January 2006): 5–32.

Johnson, Jerry, and Marty Strange. *Why Rural Matters 2007: The Realities of Rural Education Growth*. Arlington, Va.: Rural School and Community Trust, 2007.

Martin, Danny B. *Mathematics Success and Failure among African-American Youth: The Roles of Sociohistorical Context, Community Forces, School Influences, and Individual Agency.* Mahwah, N.J.: Lawrence Erlbaum, 2000.

Middleton, James A., and Photini A. Spanias. "Motivation for Achievement in Mathematics: Findings, Generalizations, and Criticisms of the Research." *Journal for Research in Mathematics Education* 30 (January 1999): 65–88.

Moses, Robert P., and Charles E. Cobb. *Radical Equations: Civil Rights from Mississippi to the Algebra Project*. Boston: Beacon, 2001.

National Council of Teachers of Mathematics (NCTM). *Principles and Standards for School Mathematics*. Reston, Va.: NCTM, 2000.

Senk, Sharon L., and Denisse R. Thompson, eds. *Standards-Based School Mathematics Curricula: What Are They? What Do Students Learn?* Mahwah, N.J.: Lawrence Erlbaum, 2003.

Stigler, James W., and James Hiebert. *The Teaching Gap: Best Ideas from the World's Teachers for Improving Education in the Classroom*. New York: Free Press, 1999.

Turner, Julianne C., and Helen Patrick. "Motivational Influences on Student Participation in Classroom Learning Activities." *Teachers College Record* 106 (September 2004): 1759–85.

Chapter 8

Factors That Motivate Aboriginal Students to Improve Their Achievement in School Mathematics

Kanwal Neel

For many years, the achievement and participation rates of Aboriginal students in mathematics in British Columbia, Canada, have been significantly lower than those of the general student population. (Aboriginal peoples in Canada comprise the First Nations, Inuit, and Métis.) The 2006 Canadian Census counted 3.8 percent of the country's total population to be of Aboriginal ancestry. This article investigates factors in community, school, and personal life that could influence the success rates of Aboriginal students in school mathematics and mathematics-related disciplines.

Devlin (2000) found many people saying they "can't do math" and compared the idea to running a marathon. He asserted that the key to being able to run a marathon (or to do mathematics) was in the individual's will. That will, motivation, or determination comes to different people at different times depending on their personal circumstances. Staying motivated about learning mathematics can be a challenge for many people, especially when students are trying to fit into multiple worldviews.

This article analyzes the transcripts of interviews with members of the Haida Role Model Program and other community members in Haida Gwaii, a remote archipelago in Northwestern British Columbia. Members of the Haida Role Model Program are of Aboriginal ancestry, belonging to the Haida First Nation, and consist of elders, professionals, and community members who go to schools and assist teachers in integrating Haida knowledge and perspectives into the school curriculum. Some community members interviewed were non-Haida but were related to the Haida nation by marriage. Interviews yielded several themes concerning how different ways of knowing affected the success of Aboriginal students in school mathematics and math-related disciplines. Interviewees focused on different aspects of the issue and offered their view from their experiences in the community. The accounts showed many connections between the types of problems that people solve in their daily lives and the concepts taught in school mathematics.

Many students find that connecting the mathematics concepts they learn in school to their daily lives is difficult. But their motivation to learn increases when they see that the mathematical skills from school are useful for their daily functioning in the home, workplace, and community (Bishop 1988). Showing Aboriginal students that traditional and contemporary cultural activities include many mathematical concepts could bring about the disposition that mathematics is useful and meaningful.

Situated Learning

When participants were asked to share a story about how they learned best, their answers invariably pointed to two factors: (1) learning was easier if it was situated in a familiar context and (2) the need to interact with those who have more experience and knowledge—such as a mentor, teacher, or a guide to mediate their learning (Vygotsky 1978)—was important to learning. Many Aboriginal students have difficulty relating to certain teachers because of cultural conflicts. They need a teacher who is a "culture broker" (Stairs 1995). A culture-broker teacher helps students move back and forth between an Aboriginal culture and the culture of Western mathematics (conventional school mathematics) and helps students deal with cultural conflicts as they arise. For nearly all students whose home worldview differs from that of school, cultural border crossing is not smooth (Aikenhead 1997).

Danny Robertson, director of the Swan Bay Rediscovery Program in Skidegate, is not Haida, but his wife, Nika Collison, is. He points out that he succeeded when the learning mattered to him and it was practical (fig. 8.1):

Fig. 8.1. Ecotourism promotes Haida Gwaii's natural and cultural environment (photo by author).

> Math was one of the subjects I found most challenging throughout elementary school and high school. . . . When I started working in the outdoor recreation business and started working as a sea kayak guide or rock climbing instructor, I really felt the pain of the lack of knowledge because simple calculations that are absolutely fundamental for safety were challenging for me. I got my unrestricted captain's ticket and I got 95 percent on the exam. It is because it was really practical and something that I could wrap my head around. It was daunting, but I was fully capable. Again, because it was relevant, there were other guides to help me, and I was interested.

Cecil Brown, a young entrepreneur, relates his experience of also having difficulty in learning mathematics, but he could learn the concepts once they were taught in smaller components and made contextual.

> I was never good at math, but I always started thinking with dollars. How could I put this money in my pocket and take it away? Fractions were the toughest thing for me, but one of my teachers just explained with fishing and measurements of a pole and it eventually all clicked in. Basically you have to break it down, break it down, break down and then after a while you start finding all these shortcuts for yourself. I had to do an upgrade in math, and this teacher really helped me because he broke it down for me. Breaking down all the concepts was like, basically, just teaching you how to add again—everything made sense.

Elder James Young (fig. 8.2) started to learn mathematics when he was only three years old. His learning took place not in a traditional way but in a situation that involved reasoning, logic, patterns, and mental calculations while he played checkers:

Fig. 8.2. Elder James Young in Skidegate, British Columbia (photo by author)

> When I was a little boy, maybe three or four years old, my uncle used to come, and every day we would play checkers. We had the checkerboard and lined

> all the checkers up—I had twelve on my side; he had four on his side. Even though he had four checkers, he used to beat me. I never gave up, and every day we played checkers. Finally, I won a game. When I won a game he added another checker to his side and played with five pieces. As I grew up and got better at the game, he had to add more checkers. Pretty soon we were playing twelve against twelve. Since the way my brain was developed, before I even started school, I was good at math.

Guujaaw, an experienced carver and president of the Haida Nation, explains that learning is lifelong, and he learned how to carve not by going to school but from others: "I learned to carve from all the peers, the old people, the old collections—that is how most of us learn, eh? We didn't go to school for that; we learned one to one . . . working as an apprentice and then learning on the job or actually doing it. It is continuous learning."

In Haida Gwaii many people used to work in fishing, forestry, or other resource-based industries, and those jobs are becoming fewer. Since many of these people lack formal education, they now need to upgrade their education to be qualified for jobs in other trades. Bobbi Parnell, who works at the Masset Learning Centre, assists clients to gain skills and education so that they become self-sufficient through employment or self-employment. She observes that many adults find it a challenge when the learning is not contextual:

> I would say there's been very little success in the Adult Basic Education that's going on in the community. Actually, working on computers doesn't work for a large number of people—they just give up. I think connecting it to their lives is really important. I think a lot of people drop out of the programs because "What is the use of this?" "What does this have to do with my life?" and so I think if they could learn math within the context of their everyday life—they're trying to train for careers they have never been in before, so they can't connect to it that way, so at least if they could connect it to their lives. I think practical projects are where they know it is a link to something—building is not cultural but practical. I have seen people who have succeeded in upgrading when they had a very specific goal.

Commonality exists among the various participants. Learning takes place in situations of coparticipation that others in that community mediate. Participants see learning not as the individual acquisition of knowledge but more as a process of social participation. Situated learning in a community of practice involves much more than the technical knowledge or skill associated with undertaking some task. Members are also involved in a set of relationships over time (Lave and Wenger 1991). Eventually "learning as internalization, learning as increasing participation in communities of practice concerns the whole person acting in the world" (Lave and Wenger 1991, p. 49).

Early Intervention and the Role of Parents

Palmantier (2005) reported that Eurocentric approaches often do not fit the needs, interests, or development and learning styles of Aboriginal students. Early intervention to address student learning difficulties in mathematics is more successful than responding later to accumulated deficits (British Columbia Ministry of Education 1999). The National Council of Teachers of Mathematics (NCTM) Position Statement on Early Childhood Mathematics Education (2007) also affirms that "high-quality, challenging, and accessible mathematics education provides early childhood learners with a vital foundation for future understanding of mathematics."

Young children are naturally inquisitive about mathematics, and teachers can build on this inquisitiveness to help students develop the positive attitude that often occurs when one understands and makes sense of a topic (Expert Panel on Early Math in Ontario 2003). "The most important connection for early mathematics development is between the intuitive, informal mathematics that students have learned through their own experiences and the mathematics they are learning in school. All other connections—between one mathematical concept and another, between different mathematics topics, between mathematics and other fields of knowledge, and between mathematics and everyday life—are supported by the link between the students' informal experiences and more-formal mathematics" (NCTM 2000, p. 132).

Beginning early, perhaps even before they are conscious of it, children form attitudes about mathematics. Often, a parent's frustration or discomfort with mathematics influences a child's perspective. The parent's attitude may well result from her or his own early mathematical experiences, forming an unbreaking cycle of negativity.

Elizabeth Moore, chief councilor for the Masset Village Council, believes in the importance of all children being numerate and that the weaker students should have mentoring opportunities. The children need to face challenging learning opportunities inside and outside the classroom. She says, "I want our children to have the needed math skills. My daughter is in grade 4, and I challenge her. The other day I asked her, 'What is 16 times 60?' I knew she didn't know the answer, but I just wanted to challenge her. She just said that the answer is big. We also need the kids to have way more opportunities on the land rather than sitting in the classroom. The kids need to be out there picking berries, gathering spruce roots, looking at the tidal pools."

Reg Davidson (fig. 8.3), a carver, feels that if as a parent he did his job, then teachers could do theirs. He also believes that people become accustomed to doing things a certain way and that changing is difficult for them. He also said that "you need to work with yourself first and then you work with your family. So you start with yourself first instead of trying to fix the world. The world will change."

Fig. 8.3. Reg Davidson starting with a scale drawing of 18:1 and using a variety of tools to carve the actual totem pole (photo by author)

While teachers and educators are still the purveyors of formal knowledge and the curriculum of mathematics education, parents and other members of society play a vital role in a child's success. Attitudes about mathematics start to form even in a child's early years. Sometimes, difficulties that students experience in school stay with them as members of society for the rest of their lives, not just through the formal years of learning. Parents need to be more involved in what is going on and understand the importance of their role in their children's education.

Changing Personal and Community Attitudes toward Mathematics

For Aboriginal students to succeed in mathematics, they and their families need to have a positive attitude and to view mathematics as a valuable component of their education. As the Royal Commission on Aboriginal Peoples reports, "Education programs, carefully designed and implemented with parental involvement, can prepare Aboriginal children to participate in two worlds with a choice of futures" (Battiste and Barman 1996, p. 442). Community attitudes toward mathematics can have a dramatic effect on students' learning of mathematics. Unfortunately, saying "Oh, I was never that good at mathematics" seems acceptable, and yet agreement that mathematics is an important school subject is almost universal (NCTM 2000). One way to stop the downward spiral of negative attitudes is to improve experiences in mathematics classrooms. Parents care a great deal about how their children feel about classroom experiences; parents do not value a classroom that bores children or makes their children feel incompetent and worthless. If student success can improve, and they begin to enjoy learning mathematics, we can begin a cycle of positive attitudes. Students will find greater success when they find greater motivation to learn mathematics. Motivation and change of attitude can improve when students do tasks that are personally relevant, explains Danny Robertson: "It is also a lot easier if kids can identify personally with what they learn—there is motivation to want to learn more about it. Like in my case, learning mathematics was brutal for me until I was a captain, and it really mattered and I really found it interesting, and then I learned what I had to do. My attitude changed."

Students often mistakenly believe that they do not need mathematics because they are "only going to be a [some occupation]," but that occupation, they find, in fact depends on mathematics. When mathematics becomes clearly important to students' futures and their daily living, they will learn better and enjoy it more at the same time. Success itself breeds enjoyment. Changing the preceding societal attitudes is a challenge, but implementing these changes would require sustained efforts from the various stakeholders.

Culturally Inclusive Pedagogy

A fundamental concept that many Aboriginal people share is *relationality*, the belief and understanding of the interconnectedness of our world and all within it. Relationality also encompasses other realities that we cannot see but are aware of (Wilson 2003). Aboriginal students participate in two cultures: that of the home and the school. Many of these students see little connection between these two cultures, and consequently many potentially rich situations from the native culture are nowhere to be found in the school (Davison 2002). James Sawyer (fig. 8.4), a carver and jewelry maker, suggests that students should get a chance to watch artists in action to learn concepts in mathematics:

Fig. 8.4. James Sawyer creates designs on tracing paper before making them into silver bracelets (photo by author).

> I think the kids would probably get a lot more interested if they were watching guys carving up poles or sculptures or even showing them the breakdown of bracelets or whatever on paintings because everything sort of starts in a

> square. From the square you go out to a three-dimensional piece—because most of the carvers will start with blocks and do everything in blocks, and then after the blocks are done and they make sure the blocks are even, then they round them into their geometric forms or ovoid forms. Everything on the poles is symmetrical—they've got a lot of mathematics and measuring they do.

Davison (2002) asserts that using cultural situations can improve Aboriginal students' mathematics learning in several ways. When the teaching of mathematics uses ideas from the culture, students value their cultural heritage more. The integration of the students' experiential mathematics with their school mathematics can help them make new connections. Guujaaw suggests that we should not just focus on the curriculum but also find a way to integrate the learning with the land (fig. 8.5):

Fig. 8.5. Is the tide coming in or going out at North Beach? (Photo by author.)

> I think that the problem with our schools here is that the real-life opportunities apparent and in the lands and culture around us are ignored. A student might learn about things all around the world, but there is little attention to seabirds, or any of the many aspects of biology, or about care of the oceans, and management about the resources which people can build careers around. The flexibility has got to be built into the curriculum to allow our kids to learn about those things that are relevant to the life in the place that we live.

Many researchers suggest that Aboriginal students' achievement improves significantly when teachers deliver culturally inclusive curricula and pedagogy in a way that accounts for learner diversity (Aikenhead 2002; Davison 2002; Hankes and Fast 2002; Irvine and Armento 2001; Jeffrey 1999; Nichol and Robinson 2000). Bishop (1988) indicates that the cultural background of students is rich in resources from which one can develop mathematics concepts. Talking about having culturally inclusive pedagogy is easy, but the reality is, if a teacher merely uses a handout or a worksheet to make a concept culturally relevant, then it defeats the purpose. The teaching needs to be imaginative and creative to motivate students to learn in ways that are interactive and engaging.

Learning on the Land

Contextualizing and making connections are powerful processes in developing mathematical understanding. "When mathematical ideas are connected to each other or to real-world phenomena, students can begin to view mathematics as useful, relevant, and integrated" (Western and Northern Canadian Protocol 2006, p. 7). About thirty years ago, the Rediscovery Program (Henley 1996) started as a single camp near Masset in the north and then grew from its successes into an international organization. The purpose of this program was to reintroduce the youth of a particular culture or nation to their culture on the lands they own. For example, every summer the Swan Bay Rediscovery Program takes place in the south part of the island, giving the youth life skills, job skills, and cultural knowledge through a variety of activities. These activities include experiences in hunting and food gathering, cedar bark weaving, and learning about historical and archaeological sites. The students learn how to read tide books, charts and mapping, compass work, and chart work. They also have to plan a meal for which they need to calculate how much food they need to bring and take the responsibility of finding resources in the wilderness. Danny Robertson, director of the Swan Bay Rediscovery Program, mentions that the staff "teach concepts that can be tied to [students'] culture—anything from just simple food gathering to safe hunting protocol to reading the tides. All the activities require a lot of problem solving and focus on mathematical skills."

Schools and Community

Many Aboriginal communities are still healing. We must understand the current realities of their communities, for the problems and challenges they face are complex (Kavanagh 2006). Many students have difficulty connecting the mathematics concepts they learn in school to what they are using in their daily lives in their communities. Mediating meaning for Aboriginal students by showing them how traditional and contemporary cultural activities have many embedded mathematical concepts could motivate them to learn. Schools need to build the confidence and trust of the community by reaching out to them and being respectful of their cultural traditions. Kim Davidson, a carpenter in Masset, wonders how to break the cycle of oppression: "In Aboriginal communities across the country, there is a system of institutionalized racism at work. What I see is that Aboriginal people that have gone through the school system have not had a very good experience. So they return to their communities with their bad experiences; their kids continue to struggle through it and they probably have bad experiences. Now, you have a whole population of people who don't understand what is going on in the world—especially with the oppression that is happening within the Aboriginal communities."

Developing an outreach program such as "Homework Helper" could help parents or grandparents understand how to help students. Like reading stories, the parents could do interactive math activities such as counting, measuring, designing, estimating, locating, or even game playing (Bishop 1988). A viable support system needs to be in place that could offer more one-on-one help for students in tutorial sessions; such sessions could occur at a venue other than the school.

Bernard Kerrigan, who formally studied education and law and now resides in Old Masset as a carver and jewelry maker, thinks that smaller classes would help. The teacher could monitor all students and ensure that they know the skills. Another factor that he thinks impedes learning is that some students miss several classes for a variety of reasons. Teachers should contact the parents or guardians right away in such a case, with the assistance of a liaison worker or the support of a home–school coordinator. This endeavor should also be part of the outreach program. He also thinks that math anxiety and a teacher's disposition toward mathematics are also a factor in students' learning:

> I had one instructor in college who loved figuring stuff out, and he just had a joy of mathematics and sees the fun—like, the way he looks at it, I couldn't imagine someone not enjoying mathematics being taught by him because he is just so enthusiastic of it. I see a lot of teachers that expect the children that have trouble with mathematics to live up to whatever expectations you have, you live up to that—so if you expect them to enjoy math and have fun with it and you teach that to them, that it is fun and not something to be afraid of.

Jason Alsop, a recent graduate, talks about the realities of a mathematics classroom. He wondered how difficult it is for a teacher to support all learners, offer differentiated instruction, and give individual attention:

> What I realized in my time is that you need the support—like, in some colleges, they have activity centers where you can go get extra help at a time when you have questions. But in a high school class, it seems like there is one teacher for the whole class and there are about four different levels for that one class. So teachers get impatient with the lower-level students at times. They don't have time to reexplain things, so they say, "I've already explained this; I don't have time to deal with you" and "I have got someone over here that is doing really well and I would rather help them do really well than help you get to the mid-level." It seems like they prioritize a little bit. They only have an hour to do their thing and teach all the students.

Today's mathematics curriculum from prekindergarten to grade 12 is usually limited to valuing aspects of the Eurocentric world by sharing its successes but ignoring those of other cultures. Students of Indigenous and multicultural heritage often face the challenge of learning in an environment that may undervalue or ignore their cultural backgrounds. *Principles and Standards for School Mathematics* (NCTM 2000) advocates the need to learn and teach mathematics as a part of cultural heritage and for life. Achieving this goal requires changing mathematics teaching approaches and introducing culturally inclusive curricula and pedagogy. Learning activities should build on a student's prior knowledge and present mathematics in an exciting and inclusive way. Context combined with content should direct teaching in the ongoing cultural quest for knowledge.

Straddling the Worldviews

Aboriginal people acquire knowledge in multiple ways, but the coming-to-know process is nevertheless extremely systematic in both Aboriginal and Western epistemologies. Vonnie Hutchingson, director of Haida Education with School District No. 50, stresses the importance of knowing that in the traditional Aboriginal worldview, everything came as a whole and there weren't discrete parts. Everything you know connected to different aspects of your life. "So, if you looked at mathematics specifically and you asked elders how it is that we use mathematics in our daily lives, they would have a really hard time trying to articulate that because it was in everything that we did and it wasn't segmented out," Hutchingson says.

Aikenhead (2002) gently warns us to proceed cautiously because misunderstanding culturally embedded meanings is easy when we do not fully share the other person's culture. He indicates that we need to respect Aboriginal knowledge and learn from the Aboriginal people and remember that gaining Aboriginal knowledge is a process of coming to know. Students need to know about different worldviews; both the Aboriginal and Eurocentric worldviews express the creative process that connects all things. Cajete (1999) explains that Indigenous peoples have historically applied the thought process of creative science within cultural contexts, which are holistic. Every Indigenous culture has an orientation to learning that is metaphorically represented in its art forms, its way of community, its language, and its way of understanding itself in relationship to its natural environment.

Language is said to be the root of many cultures. Words in a language, the ideas and feelings that they represent, and how they are spoken allow people to fully express their traditional beliefs (Reed 1999). The Haida culture is no different. Diane Brown, a Haida language teacher, expresses the difficulty that students face in learning their heritage language: "It is hard for the kids to be good in both cultures. I think right now they are more in the white world than our world. I think there is an inner need and an inner will to be that, and some of them will come here and think they could learn the language just like that, and the reality is it's difficult, so they don't come very often."

Shared Learnings (British Columbia Ministry of Education 1998) states that Canada's Aboriginal peoples value a legacy of oral tradition that accounts for each group's origins, history, and spirituality. Stories bind a community with its past and future, and oral traditions reach across generations, from elder to child, and bear witness to the creation of women and men and how they populated the land. A need is apparent to recognize the existence of multiple worldviews and knowledge systems and to find ways to understand and relate to the world in its multiple dimensions and varied perspectives.

Conclusions

The older generation is determined to improve the educational success of their children. The definition of success is complex and unique to each individual and community. Should success be based on the scores of a large scale, or should it "involve children being self-confident, understanding their own culture and traditional values, and have a positive self-identity" (Kavanagh 2006, p. 20)? According to the British Columbia Ministry of Education (2006), data have consistently shown that many of the province's Aboriginal students perform lower than non-Aboriginal students in their achievement in mathematics. This trend has become a cause for concern because success is measured solely on the basis of data from provincial assessment. Such results do not reveal the whole picture of the community. Many members of the community are adamant that their children should learn mathematics that is "authentic." They also want to see that the curriculum acknowledges and respects their culture, but they do not want a watered-down curriculum for their children. Success is a difficult concept that should not be based only on academics or economics but rather should look at life in balance.

The complexities that come into play when two fundamentally different worldviews converge present a great challenge. Many participants suggest that Aboriginal students should seek knowledge from both worldviews and recognize their interconnectedness. Rather than a view that the problem is to get Aboriginal people to buy into the modern system, we need a dialogue that will involve both cultures. Aboriginal people may need to understand mainstream mathematics, but not at the expense of what they already know. Non-Aboriginal people, too, must recognize the existence of multiple worldviews and knowledge systems and find ways to understand and relate to the world in its multiple dimensions and varied perspectives. Interconnected ways of knowing can motivate students in learning mathematics as well as increase all students' levels of mathematics.

Several themes have emerged for what can be useful in integrating students' experiential mathematics with their school mathematics, to help motivate them and make new connections to improve achievement. Learning needs to be situated and in context. Students need to interact with those who have more experience and knowledge, such as a mentor, teacher, elder, or a guide, to mediate their learning. Students need to be able to straddle different worldviews imaginatively and creatively. Early intervention with increased parental involvement can make a difference to students who have difficulties in learning mathematics; this approach is far more effective than responding later to accumulated deficits. Changing societal attitudes toward mathematics can be challenging, but doing so can have a dramatic effect on the learning of mathematics by students. Where possible, teachers should use culturally inclusive pedagogy that accounts for learner diversity. Schools need to change the structure inside and outside the classrooms to create more outreach programs that use the resources of the land and the expertise of role models. Implementing changes is not easy. For a variety of reasons, talking about these ideas and themes is easier than implementing them. Progress will be slow, but first we need to take steps with the intention of ongoing improvement.

REFERENCES

Aikenhead, Glen S. "Toward a First Nations Cross-Cultural Science and Technology Curriculum." *Science Education* 81 (April 1997): 217–38.

———. "Cross-Cultural Science Teaching: Rekindling Traditions for Aboriginal Students." *Canadian Journal of Science, Mathematics, and Technology Education* 2 (July 2002): 287–304.

Battiste, Marie, and Jean Barman, eds. *Gathering Strength: Report of the Royal Commission on Aboriginal Peoples*, vol. 3. Ottawa, Ont.: Canada Communications Group Publishing, 1996.

Bishop, Alan. *Mathematical Enculturation.* London: Kluwer Academic Publishers, 1988.

British Columbia Ministry of Education. *Shared Learnings: Integrating BC Aboriginal Content K–10.* Victoria, B.C.: Ministry of Education, 1998.

———. *Report of the Mathematics Task Force.* Victoria, B.C.: Ministry of Education, 1999.

———. *Aboriginal Report—2005/06: How Are We Doing?* Victoria, B.C.: Ministry of Education, 2006.

Cajete, Gregory. *Igniting the Sparkle: An Indigenous Science Education Model.* Skyland, N.C.: Kivaki Press, 1999.

Davison, David. "Teaching Mathematics to American Indian Students: A Cultural Approach." In *Perspectives on Indigenous People of North America*, edited by Judith Hankes and Gerald Fast, pp. 19–24. Reston, Va.: National Council of Teachers of Mathematics, 2002.

Devlin, Keith. *The Math Gene. How Mathematical Thinking Evolved and Why Numbers Are Like Gossip.* New York: Basic Books, 2000.

Expert Panel on Early Math in Ontario. *Early Math Strategy.* Toronto, Ont.: Ontario Ministry of Education, 2003.

Hankes, Judith, and Gerald Fast, eds. *Perspectives on Indigenous People of North America.* Reston, Va.: National Council of Teachers of Mathematics, 2002.

Henley, Thom. *Rediscovery: Ancient Pathways, New Directions*. Edmonton, Alta.: Lone Pine Publishing, 1996.

Irvine, Jacqueline, and Beverly Armento. *Culturally Responsive Teaching.* New York: McGraw–Hill, 2001.

Jeffrey, Deborah. *Task Force on First Nations Education*. Vancouver, B.C.: British Columbia Teachers' Federation, 1999.

Kavanagh, Barbara. *Teaching in a First Nations School.* W. Vancouver, B.C.: First Nations Schools Association, 2006.

Lave, Jean, and Etienne Wenger. *Situated Learning: Legitimate Peripheral Participation.* Cambridge, UK: Cambridge University Press, 1991.

National Council of Teachers of Mathematics. *Principles and Standards for School Mathematics*. Reston, Va.: National Council of Teachers of Mathematics, 2000.

———. "What Is Important in Early Childhood Mathematics? A Position of the National Council of Teachers of Mathematics." 2007. www.nctm.org/about/content.aspx?id=12590 (accessed June 7, 2010).

Nichol, Ray, and Jim Robinson. "Pedagogical Challenges in Making Mathematics Relevant for Indigenous Australians." In *International Journal of Mathematics Education in Science and Technology* 31 (July–August 2000): 495–504.

Palmantier, Monty. "Building a Community of Communities: Results and Discussion of National Roundtable on Aboriginal ECD: What Can Research Offer Aboriginal Head Start?" 2005. www.coespecialneeds.ca/PDF/ahsroundtable.pdf (accessed June 22, 2005).

Reed, Kevin. *Aboriginal Peoples: Building for the Future*. Don Mills (Toronto), Ont.: Oxford University Press, 1999.

Stairs, Arlene. "Learning Processes and Teaching Roles in Native Education: Cultural Base and Cultural Brokerage." In *Gathering Strength: Report of the Royal Commission on Aboriginal Peoples*, vol. 3., edited by Marie Battiste and Jean Barman, pp. 139–53. Vancouver, B.C.: UBC Press, 1995.

Vygotsky, Lev S. *Mind in Society: The Development of the Higher Psychological Processes*. London: Harvard University Press, 1978.

Western and Northern Canadian Protocol. *The Common Curriculum Framework for K–9 Mathematics.* Alberta: Alberta Education, 2006.

Wilson, Shawn. "Progressing toward an Indigenous Research Paradigm in Canada and Australia." *Canadian Journal of Native Education* 27, no. 2 (2003): 161–78.

Chapter 9

What Motivates Mathematically Talented Young Women?

Hortensia Soto-Johnson
Cathleen Craviotto
Frieda Parker

Andreescu and colleagues (2008) researched students who perform well on mathematical competitions, such as the International Mathematical Olympiad, the Putnam, the American Mathematics Competitions, and MATHCOUNTS. The authors found minimal diversity among the few participating women. Typically, Asian and Eastern European women partake in these events but are not top scorers. Andreescu and colleagues (2008) believe that sociocultural, educational, or environmental factors may explain why some young women are more apt to participate in mathematics competitions. In our study, we investigated factors that contribute to young women's intentions to pursue mathematics and to their disposition toward mathematics. We took a phenomenological approach to the research (Patton 2002). We used a sample of participants who attended a camp for mathematically talented young women. Our research questions included the following:

1. What motivates camp participants to pursue the study of mathematics?
2. What is the disposition of camp participants toward the study of mathematics, and what contributes to their disposition?
3. What environmental factors influence participants to pursue mathematics?

Background on Motivation

Schunk (2004) defines motivation as "the process of instigating and sustaining goal directed behavior" (p. 329) and gives factors (fig. 9.1) that contribute to motivation. Schunk's internal sources of motivation fall into *Adding It Up*'s definition of disposition: "the tendency to see sense in mathematics, to perceive it as both useful and worthwhile, to believe that steady effort in learning mathematics pays off, and to see oneself as an effective learner and doer of mathematics" (Kilpatrick, Swafford, and Findell 2001, p. 131).

Internal Factors	External Factors
Personal goals	Peers
Values	Family
Needs	Instructional variables
Persistence	Teacher
Effort	Feedback
Affective variables	Environment

Fig. 9.1. Motivation factors

Attribution theory and *self-efficacy beliefs* are two recurring aspects of motivation and disposition in literature related to gender and mathematics. Thus, we focus the literature on these two facets. Attribution theory explains what people attribute their successes and failures to and predicts achievement on the basis of attribution. Schunk (2004) defines self-efficacy as “personal beliefs about one’s capabilities to learn or perform actions at designated levels” (p. 112).

Lloyd, Walsh, and Yailagh (2005) attempted to determine whether recent academic gains of young women in mathematics are associated with heightened perceptions of self-efficacy and performance attributions. Lloyd and colleagues documented how boys tend to ascribe their mathematics successes to internal factors and their failures to external factors. They also cited research indicating that boys generally have a higher self-efficacy than girls do. In an effort to address their investigation, the researchers obtained data in the form of mathematics test scores and course grades as pre- and postachievement measures from fourth- and seventh-grade students in British Columbia. They gathered success and failure attribution data from a survey. Students attributed success to effort, strategy, ability, teacher’s help, help from others, and task ease. Lack of these same traits contributed to the failure attributions. The researchers created a rating scale to measure students’ self-efficacy on four categories: numbers, patterns and relationships, shape and space, and statistics and probability.

Lloyd and colleagues (2005) found no differences between achievement of girls and boys on the pretest, but at the end of the course, the girls’ grades were higher than the boys’ grades. The researchers did not find statistically significant differences on success attribution by gender. Both sexes attributed their success to effort equally and rated ability as most important in explaining their successes. Girls and boys had almost identical attributions for failure, but the girls were more likely to blame a lack of teachers’ help for their failure. This study showed that despite girls' achievement meeting or exceeding the boys', girls were less confident than boys. The authors suggest that "students whose self-efficacy is higher or lower relative to their actual achievement show poorer achievement than those students whose self-efficacy is commensurate with their actual achievement" (Lloyd et al. 2005, p. 403). Therefore, some environments do not maximize girls' potential, although the girls' achievement is strong. The study found parallel results for elementary and middle school students but did not measure the attribution and self-efficacy of high school students. Our study investigates these facets in high school girls through journals and interviews.

Pajares's (2002) research shows that girls (especially at the elementary and middle school level) generally employ effective methods to learn and that they express confidence in their methods. Pajares suggests that students' confidence influences their academic motivation and their academic success and that students' academic confidence differs by gender. Pajares recommends that teachers give appropriate feedback, goal setting, and modeling for students to promote their self-efficacy. Pajares explains that peers are also models, and therefore "it is important to select peers for classroom models judiciously so as to ensure that students view themselves as comparable in learning ability to models" (p. 122).

Middleton and Spanias (1999) offer a theoretical overview of the interplay between motivation and achievement in mathematics. The authors argue that the research on motivation is not consistent but mention some common themes in the literature:

- Students' perceptions of success in mathematics are influential in forming their motivational attitudes.
- Although motivation toward mathematics develops early and maintains stability over time, teachers' actions and attitudes influence its development.
- Students need opportunities to develop intrinsic motivation in mathematics, which is generally superior to giving extrinsic incentives for achievement. Offering such opportunities may require teachers to modify how they teach mathematics courses since careful instruction affects motivation.
- External factors such as gender role stereotyping, teacher expectations, and peer pressure may prompt girls to be less motivated to pursue more mathematics. For example, teachers' thoughts and behaviors may "undermine their students' achievement motivation by reinforcing failure-oriented attributions, especially for their female students" (p. 72).

Camp Description

The participants of this study attended a mathematics camp. Las Chicas de Matemáticas: University of Northern Colorado Mathematics Camp for Young Women was a one-week residential camp for thirty mathematically talented young women who had completed grades 8–11. The institution's office of enrollment management and the Mathematical Association of America Tensor Foundation funded this program. Figure 9.2 shows the goals of the camp. We selected the camp participants on the basis of their high school mathematics grades, a letter of support from their mathematics teacher, and a statement indicating why they wanted to attend the camp.

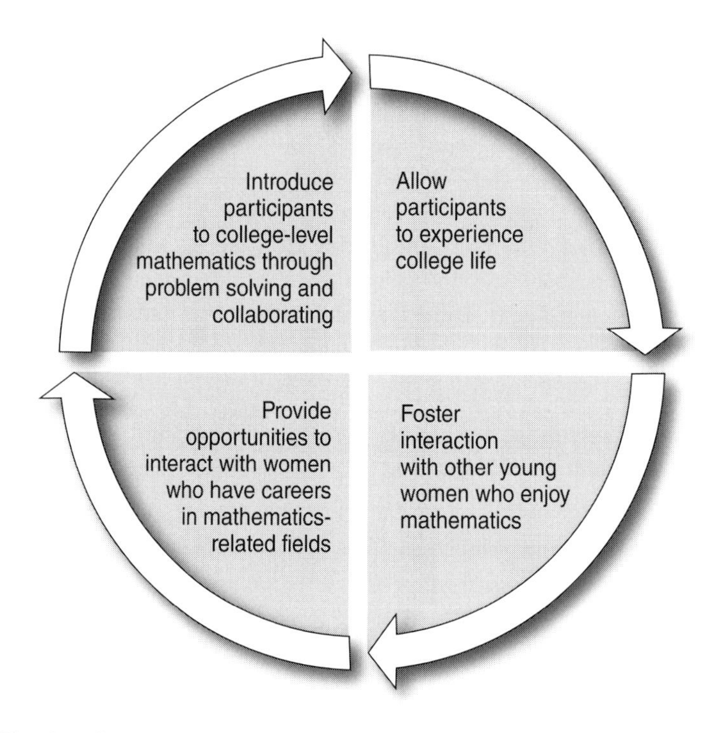

Fig. 9.2. Goals of Las Chicas de Matemáticas: University of Northern Colorado Mathematics Camp for Young Women

The camp participants learned about mathematical modeling and number theory. In the modeling class, they explored properties of binomial coefficients by using geometrical methods (counting the number of pathways on a lattice grid) and reviewed properties of the number *e*. They applied this information to analyze some classical and modern problems in probability involving repeated sampling without replacement, such as the birthday problem. This task entails determining the probability that in a set of *n* randomly chosen people, some pair will have the same birthday. Other classroom simulations included the game of repeatedly withdrawing balls from an urn until someone draws a black ball, as well as a Web-based encryption problem: deciding how long a randomly created ID number must be for one to be reasonably confident that no two visitors to a Web site accidentally obtain the same ID numbers.

In the number theory course, the participants investigated topics leading to elementary coding theory. They applied the Euclidean algorithm, which assists with determining the greatest common divisor; the Fundamental Theorem of Arithmetic (the prime factorization of every integer greater than 1 is unique); and clock arithmetic to generalized Caesar ciphers and exponentiation ciphers. For example, with the Caesar ciphers, the students used clock arithmetic to code and decode messages where each letter of the alphabet is replaced with the letter k positions after the letter in the plaintext message. If $k = 13$, then one would code the letter t (the twentieth letter in the alphabet) in a message by using the function $f(t) = (t + 13) \bmod (26)$. Thus, the letter t would be replaced with the seventh letter in the alphabet since $7 = 33 \bmod (26)$, and the word math was coded as "zngu."

Four female preservice secondary mathematics teachers served as counselors. They lived in the dorms with the participants, offered help in the classroom, assisted with study sessions, and provided general supervision. Female guest speakers visited every day to share how they used mathematics in their careers. In the evening, the participants socialized to balance the day.

Camp Participants

From surnames and discussions with the young women, we found that eighteen participants were Hispanic, 10 participants were Anglo, one participant was Taiwanese, and another was from India. All participants completed journal entries responding to prompts related to their motivation and disposition toward mathematics, and they filled out an exit survey. The journals had a twofold purpose. First, they served as a venue to communicate with all the camp participants about their experience with the camp. Second, they gave us ideas for interview questions. Nine of our camp participants participated in an interview. We selected eight of the nine participants before the camp on the basis of their ethnicity and grade level. Although we did not intend to select another participant, we included a ninth interviewee because of her classroom contributions and journal entries. Our research participants were diverse in their ethnic backgrounds and grade levels.

Information Gathered

In their journals, our camp participants responded to prompts such as those in figure 9.3. Two researchers, who were also camp faculty, read and commented on the participants' journal entries every day. In an exit survey, students responded to questions regarding the structure of the camp and described how the camp influenced them most. During the interviews, the participants shared their mathematics autobiography and responded to inquiries related to our research questions. Although we collected journals and exit surveys from all students, we used only the data from our nine interviewees.

1. How would you describe your relationship with mathematics?
2. What goals do you have related to mathematics?
3. What did you find interesting about our engineer guest speakers' presentations?
4. Have your experiences at the Mathematics Camp affected your confidence or motivation as a mathematics learner? Explain.
5. What motivates you to work hard to learn mathematics?
6. How do the people in your life, such as your friends, classmates, teachers, parents, etc., affect your motivation to learn mathematics?
7. How do they affect your confidence in your ability to learn mathematics?

Fig. 9.3. Interview questions

Results

In this section, we address each of our research questions regarding motivation, disposition, and environmental factors. We incorporate student quotes to support our assertions. All names are pseudonyms, and we use *i, j,* or *s* to indicate whether the quote appeared in an interview, journal, or survey, respectively.

Motivation

Several factors appear to motivate our participants to pursue mathematics. We categorized the motivational factors into two umbrella terms: academic performance and future goals.

Academic performance

Two themes emerged for academic performance: acceleration and grades. Our participants reported that acceleration into higher-level mathematics classes had a positive impact on their motivation. This acceleration sometimes meant advancing to a mathematics class that was one or two grade levels ahead on the basis of their classroom performance or standardized test scores. The young women perceived that the challenging mathematics courses and the opportunity to be in an accelerated class motivated them to perform well.

This advancement caused several students to try harder in their mathematics courses. During her interview, Gail reported that when she was "bumped up" in middle school to prealgebra she started giving mathematics "her all" and putting more time and energy into it. Initially, the class was difficult, but the teacher made her respect mathematics, and it motivated her to become more diligent.

Gail (i): My prealgebra teacher was really hard, and he made me respect mathematics a lot more. I like the challenge of mathematics.

Although many participants were accelerated during fifth grade, Caroline moved to an advanced class her freshman year. Motivated by her love for the subject, she chose to join a science, mathematics, and technology magnet program. Although she worried that the courses might be too difficult, she was determined to do well.

Caroline (i): This was my chance, so I had to do it or else I would regret it. I would rather do it [and] not do well than regret not doing it and not have a chance again.

This opportunity for acceleration motivated her to make a strong attempt to succeed in her mathematics courses, and her persistence paid off.

Some students found mathematics to be fun when they had the opportunity to be in advanced classes and study challenging mathematics. In sixth grade, Faith's teacher advanced her by two years, and she became an average, good student instead of the top student in her class. For her, the challenge of the course made mathematics enjoyable.

Faith (i): If I were in the regular math class it would be like doing like 2 + 2, and that wouldn't be fun—not literally, but that is what it would be like for me. It is the challenge that makes it fun.

Several young women like Faith reported that the challenge of mathematics motivates them to do well.

Grades strongly motivated many of our participants. Some of these students had a minimum grade they would accept. Talisa did not particularly like mathematics, but she was motivated to earn a grade of B+ or better.

> *Talisa (j):* What motivates me is that if I don't work hard to learn mathematics, I won't get good marks, and if I don't learn, finding a good job will be hard.

Similarly, Diana said she was disappointed in herself if her grades dropped, and she quickly sought help from her teacher if she got a C. For some students, grades indicated that they had learned as much as they possibly could, which also motivated them. Grace tried demanding problems because she "wanted the extra points to get that A+; the A wasn't good enough." She committed extra time to earn straight As.

Future goals

Many participants stated that they were motivated to do well in mathematics to accomplish their career goals and to be able to help others. Several participants would be first-generation college students and aspired to pursue degrees that demand the study of mathematics. Felicity is unsure of her career goals, but she nonetheless intends to continue to study mathematics in college because she believes that it is important for all her career choices. Gail's desire to obtain the most prestigious scholarship in the state motivates her to obtain high grades. Caroline is motivated to do well in part because she wants the American dream: "home, car, the good life." She is inspired to work hard and expressed confidence in her ability to succeed in college.

Several students mentioned their motivation to help others through their career choices. Talisa wants to understand mathematics to help her future children with their mathematics homework. Grace aspires to become a lawyer and help others, but she is considering becoming a mathematics teacher because mathematics teachers influence students' lives. Several young women were interested in medical careers because they wanted to help others; they perceived that they must do well in mathematics to do so.

Disposition

Overall, our participants' dispositions toward mathematics were positive, but these dispositions did not come easily or quickly. Experiences over time molded their dispositions. Most participants expressed confidence in their abilities to succeed in mathematics, as Nicole's following comments indicate. Our participants' confidence in mathematics may have contributed to their positive outlooks toward mathematics. Students attributed liking mathematics to the internal consistency of mathematics, the challenge of mathematics, and the relationships with their teachers.

Nicole (i, j): I know once I get it, it's there and it's not going to go away. I know I am good at it, so I will be able to help a lot, and I am also really good at explaining it.

Internal consistency of mathematics

Some participants, such as Nicole, commented that they liked that mathematics problems have only one answer. Participants said that they liked mathematics because it has set procedures to find the unique answer to a question. Another student, Wendy, enjoys how conceptual understanding of mathematics allows her to solve other mathematics problems.

Nicole (i): I just really like the consistency of math. I like the steps and ordered process of math. So, you have the format, and if there are exceptions you know the format, and you can still get the answer.

Wendy (i): My teacher taught us differently than I had learned before . . . we get to really think about it and understand it . . . and it seems more interesting. I see how it works and never forget it.

Challenge of mathematics

The challenge of mathematics contributes to our participants' motivation and to their dispositions toward mathematics. Several participants enjoy the challenge of mathematics and believe that this is what makes mathematics fun. Some participants recognize that their accelerated class challenged them, but they welcomed this opportunity. Some participants describe mathematics as a puzzle and discuss their persistence at solving a problem.

Teacher relationships

Students also credit their teachers for their dispositions toward mathematics. Whereas some students describe the content as challenging, others recognize how the teacher contributed to developing a challenging course. Still others, such as Felicity, comment about how their teachers engaged them with the mathematics, which helped the students remember how much they enjoyed mathematics.

Felicity (i): I like teachers who get the students involved and let students help one another. When my teacher did this, I remembered how much I liked math.

Environmental Factors

Personal goals, an inner drive, and individual needs greatly influenced our participants' motivations and dispositions toward mathematics, but external factors also contributed. Most of the young women remarked how relationships with their teacher, peers, and family motivate them and affect their dispositions toward mathematics. They also commented on how the mathematics camp sculpted their confidence in mathematics and influenced their new career goals.

Teachers

Many of the young women spoke of the importance of one-on-one time with their teachers. Some students felt that they understood the subject better if they worked alone with the teacher. Others were willing to go in after school to discuss mathematics with the teacher because of the one-on-one relationship they had with the teacher and because the teacher encouraged questions. Our participants feel motivated because their teachers noticed their potential in mathematics and believed in them.

Our participants also believed that their teachers built up their confidence in mathematics. They credited their interest and inspiration in mathematics to their teachers' support and encouragement. Faith described how her teacher demonstrated belief in her by persuading Faith to participate in other activities such as MATHCOUNTS and our camp.

Faith (i, j): My teachers encourage me to do well in school and encouraged me to come to math camp. My teacher has really been the one introducing me to math competitions.

Peers

Our participants remarked on how their peers affected their motivations and dispositions toward mathematics. Most comments centered on sharing knowledge through group work, especially when they feel safe not knowing the answers and asking questions. One student, whose primary language is not English, reported wanting to speak English better so that she could share her mathematics skills with her classmates. Some participants, such as Gail, described how working with other students of the same caliber and who had the same goals also contributed to their interest and work ethic in mathematics.

Gail (i): In this class [honors geometry], people are more passionate about the class and their grades, so that's what I do. That's how I set my example—they set the example, and I am more of a follower, so I started doing better.

Family

The participants' families were a primary support system. Faith spoke of her father, who could help her in mathematics, and the special "math bond" she shared with her sister. For fun, she and her younger sister create and solve mathematics problems. Caroline described how her parents value education, point out role models, and encourage her enthusiastically. She attributed part of her success in mathematics to the high expectations that her parents instilled in her. Diana summed up the importance of relationships as a motivating factor.

Caroline (i): It's hard to live up to, but it is my habit now—they don't have to ask for it because I want it.

Diana (j): What motivates me to learn mathematics is my parents, instructors, close friends, and most importantly myself in pursuing these goals.

Not all our participants had such support at home. Nicole perceived her mother as uninterested and unsupportive. Nicole's desire for support inspired her to create a support "family" from her church, and one of her "church moms" brought her to mathematics camp.

Math camp

Our participants' experience at mathematics camp continued to shape motivation and disposition toward mathematics. Most of our participants began to consider other career choices. Although the students did not feel confident on the first day of camp, the mathematics camp contributed to their confidence. Diana's comments illustrate the transformation in her confidence during camp:

Diana (j): I'm not too confident about me feeling prepared for today's class regarding my math background.

Diana (i): My experience at math camp has affected my confidence. I feel like I am a better learner here than in school. Knowing that I can do well here makes me confident to take on any other obstacles I have in the future.

Diana (s): I've been more confident, and I am even more interested in math than I ever was.

Conclusions and Suggestions

Our research illustrates how motivation, disposition, and environmental factors contribute to the pursuit of mathematics. It indicates the interplay between our participants' motivation to do mathematics and their disposition toward mathematics. It also illustrates how environmental factors influence motivation and disposition. Our research extends the work of Schunk (2004) and Pajares (2002), suggesting that motivation also affects disposition in mathematics. For example, our participants expressed confidence about their ability to succeed in mathematics because their teachers took the initiative to accelerate them into higher-level mathematics classes. The challenging mathematics classes also contributed to their positive disposition toward mathematics. Disposition toward mathematics appears to be dynamic rather than static.

Andreescu and colleagues (2008) hypothesized that environmental factors contribute to whether young women participate in mathematical competitions. Our research builds on this hypothesis by identifying two environmental factors that contribute to young women's motivation and disposition. Our participants' relationships with their parents, teachers, and peers contributed to their desire to engage in challenging mathematics. These relationships also fostered support and encouragement, which helped mold our participants' confidence in mathematics. Participation in extracurricular activities, such as our mathematics camp or various mathematics competitions, also fashioned their motivation and disposition. Several participants attributed their success in mathematics to their teachers, corroborating the results of Lloyd and colleagues (2005). Many of our participants were first-generation students who sought educational guidance from their teachers, which may explain our findings.

The results of this research also have implications for practitioners. Teachers must recognize talented young women in mathematics and engage them in challenging mathematics at an early age. Many participants remarked that they did not recognize their mathematical abilities until an elementary, middle school, or high school teacher pointed these abilities out. Acknowledging students' mathematical talents can lead to accelerating students into a higher-level mathematics course or encouraging their participation in extracurricular mathematics activities. This responsibility lies in the hands of both teachers and counselors, who can direct students to math camps, competitions, and other opportunities or who can invite inspirational speakers. Mathematics teachers at all levels should create learning environments where students can collaborate with one another and learn mathematics conceptually.

Motivation and disposition toward mathematics, as well as environmental factors, influence whether mathematically talented young women pursue mathematics (fig. 9.4).

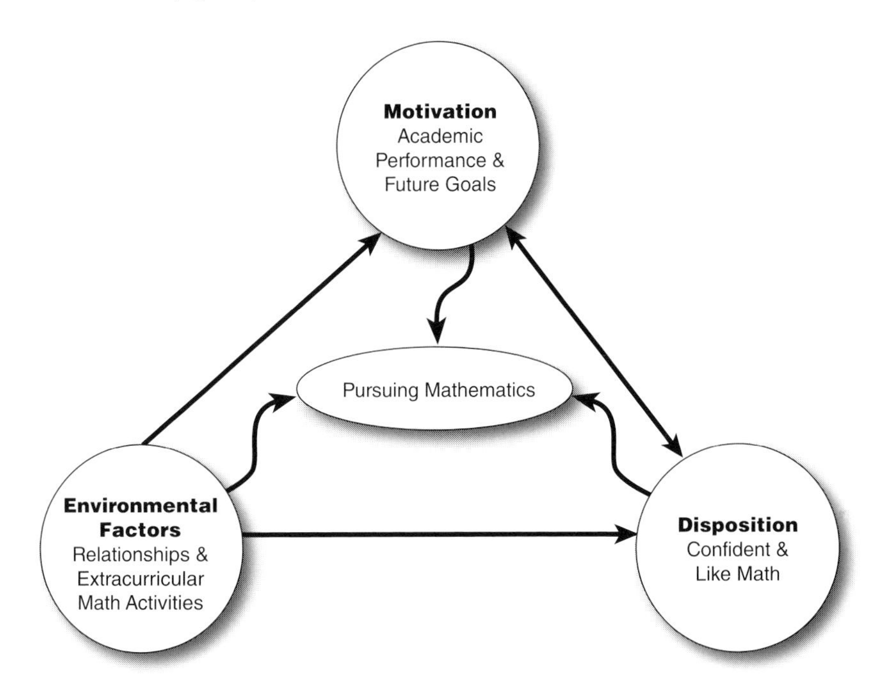

Fig. 9.4. Models for pursuing mathematics

Figure 9.4 depicts how motivation and disposition shape one another since one's views of mathematics influence career goals and vice versa. However, environmental factors, such as personal relationships, can motivate and affect one's disposition toward mathematics.

REFERENCES

Andreescu, Titu, Joseph A. Gallian, Joseph M. Kane, and Janet E. Mertz. "Cross-Cultural Analysis of Students with Exceptional Talent in Mathematical Problem Solving." *Notices of the American Mathematical Society* 55 (November 2008): 1248–60.

Kilpatrick, Jeremy, Jane Swafford, and Bradford Findell, eds. *Adding It Up: Helping Children Learn Mathematics.* Washington, D.C.: National Academies Press, 2001.

Lloyd, Jennifer E. V., John Walsh, and Manizeh Shehni Yailagh. "Sex Differences in Performance Attributions, Self-Efficacy, and Achievement in Mathematics: If I'm So Smart, Why Don't I Know It?" *Canadian Journal of Education* 28, no. 3 (2005): 384–408.

Middleton, James A., and Photini A. Spanias. "Motivation for Achievement in Mathematics: Findings, Generalizations, and Criticisms of the Research." *Journal for Research in Mathematics Education* 30 (January 1999): 65–88.

Pajares, Frank. "Gender and Perceived Self-Efficacy in Self-Regulated Learning." *Theory into Practice* 41 (Spring 2002): 116–225.

Patton, Michael Q. *Qualitative Research and Evaluation Methods.* Thousand Oaks, Calif.: Sage, 2002.

Schunk, Dale H. *Learning Theories: An Educational Perspective*. 4th ed. Upper Saddle River, N.J.: Pearson, 2004.

Part III

Motivation and Disposition in the Classroom

PART III of this yearbook is titled "Motivation and Disposition in the Classroom." The seven chapters in this section cover diverse topics that emphasize connections to classroom practices.

In "Metaphors: A Powerful Means for Assessing Students' Mathematical Disposition," the authors introduce and discuss how responding to a short "math metaphor" can be a vehicle for assessing students' dispositions. The chapter gives examples of metaphors that show how and why students like and dislike mathematics. The authors offer two methods for analyzing metaphors, give practical tips for using metaphors to assess dispositions, and discuss the benefits when teachers build on the data to develop strategies to better engage students.

The authors of "Using Prediction to Motivate Personal Investment in Problem Solving" discuss an instructional strategy using prediction that can support opportunities for active learning and foster personal investment by students to increase motivation. Although the example comes from work with students in the middle grades, the idea is applicable across grade levels. Authors describe using prediction to preview mathematical ideas of a lesson and activate prior knowledge to anticipate plausible results. They suggest that discussing the predictions focuses students' attention on mathematical meaning and structure of problems and leads to authentic explorations.

"Developing Persistent and Flexible Problem Solvers with a Growth Mindset" examines a situation where both teachers and students engaged in activities that developed persistence and motivation. The research lessons developed in a Lesson Study format, while focusing on student work, document teachers' efforts to build flexibility based on a growth mindset. The chapter discusses features of research lessons that elicited student's persistence and motivation during problem solving. The authors also describe and illustrate core instructional practices at several grade levels for developing persistent and flexible problem solvers.

If you have taught for any length of time, you will see one of your own students in the discussion of productive disposition in "Mark's Development of Productive Disposition and Motivation." As you read about Mark, you can sense his excitement as he comes to class bubbling with excitement, ready to discuss insights on a problem that had previously puzzled him. The authors use Mark's story to emphasize the importance of self-efficacy, productive disposition, and the role of agency and intellectual need, as well as how these developed over time for Mark.

"Listening to Mathematics Students' Voices to Assess and Build on Their Motivation: Learning in Groups" discusses an investigation of sixth-grade students' motivation to participate in small groups. The author emphasizes the decision to develop a focal classroom activity that would support mathematics learning while also engaging students. Concerning group work, the authors examine students' examples related to learning mathematics, developing social skills, developing autonomy, and the costs associated with small group work. They also point out that teachers are not always proficient at predicting what motivates their students, so the effort to gather data from students related to their motivation is an important, productive venture.

In "Motivating Mathematics Students with Manipulatives: Using Self-Determination Theory to Intrinsically Motivate Students," the authors distinguish between intrinsic and extrinsic motivation and outline why teachers should strive to intrinsically motivate their students. When the authors relate intrinsic motivation to novelty, challenges, exploring, and learning, you can see why they connect this concept to high-quality learning and creativity. They poignantly remind us that extrinsic rewards decrease intrinsic motivation. The authors explore two vignettes related to using manipulatives for learning mathematics, and they discuss the teachers' role in instruction. They conclude that how—not whether—teachers use manipulatives is most important to intrinsically motivate students and that teacher actions should support students' autonomy, competence, and relatedness.

The author of "Using Movies and Television Shows as a Mathematics Motivator" offers many examples of mathematics in the entertainment media that can appeal to students at various levels. Among the examples are arithmetic puzzles involving the Fibonacci sequence, number theory, coding theory, geometry, and algebra. Examples come from the movie *The Da Vinci Code* and the television program *Futurama*. Mathematics aficionados will happily recall the classic "water in jugs" problem from *Die Hard: With a Vengeance* and will probably proceed to solve it. The Monte Hall problem will remind readers that a good mathematics problem can engage people of all ages and abilities as the author explores how solving this seemingly simple problem caused mathematicians to disagree.

As you read the articles in this section, you might consider some of these questions:

- What topics besides food, color, or animals could you consider using with your students when exploring a math metaphor? Why would these be good choices for the level of students you teach? Would you consider sharing the various metaphors that students write with the rest of the class?
- How can using prediction motivate students? How can a teacher change a skill-based lesson to a problem-focused lesson so that students have to make a prediction at the beginning of the lesson?
- Why is the message of "The Little Engine That Could" important for teachers to remember? What instructional strategies can you use to help students develop the belief that they can succeed?

- How can we balance the need to scaffold lessons with the need to engage children in problem-solving experiences where struggling to understand a problem is an important consideration?
- Why is choosing a good learning task important when asking students to work in groups? Why is group discussion focusing on ideas that are not yet solidified as important as discussing the solution to an already-solved problem?
- What is one question related to motivation that you think you know how your students will respond to, but you have not yet collected the data? When will you plan to ask this of your students, and how will you use the information you obtain?
- What example of mathematics in media will you consider using? Why is bringing examples of mathematics from the current culture to the classroom important?

Melfried Olson

Chapter 10

Metaphors: A Powerful Means for Assessing Students' Mathematical Disposition

Jinfa Cai
F. Joseph Merlino

ALL TOO OFTEN, a student will attempt to solve a challenging mathematics problem but will quickly give up, saying, "I can't do it." Students' despair over their supposed inability to do mathematics does not mean that they can't solve new problems. Instead, it usually means that students do not have the disposition or learning habits that enable them to persevere in working through challenging problems without constant teacher help (National Council of Teachers of Mathematics [NCTM] 1989). Mathematics teachers have a near-universal lament of "low student motivation"—even for their more able students. Although teachers are keenly aware of the degree to which their students are attentive and engaged in their class, or not, teachers often face challenges on how to improve student attitudes and inspire a love and appreciation of mathematics. To devise strategies for increasing student motivation in their classrooms, teachers must get to the heart of their students' dispositions toward mathematics.

In this article, we give research-based insights into how teachers can use a short metaphor survey instrument to reliably assess students' mathematical dispositions and map the emotional terrain of each classroom regarding students' assent to being instructed. We use the term *dispositions* to include not only attitudes toward mathematics but also how a student thinks and acts when learning mathematics (NCTM 1989). Both attitudes and thought processes are hard to capture.

In everyday life, people routinely use metaphors to describe complex concepts and to express their feelings and thinking. A striking metaphor can often result in deeper understanding and reveal underlying meanings otherwise difficult to grasp (Lakoff and Johnson 1980).

Mathematics education researchers have used metaphors to understand teachers' perceptions of mathematics and the teaching of mathematics (Bullough 1991; Cooney et al. 1985; English 1997; Merlino 2001; Miller and Fredericks

1988; Munby 1986; Presmeg 1992; Sfard 1998; Wolodko, Willson, and Johnson 2003). We found that asking students to describe their mathematics experiences by using food, color, and animal metaphors reveals much about their dispositions toward mathematics. Using metaphors affords a much richer context and vocabulary to express and communicate students' dispositions than multiple choice–based item response surveys and offers a means to reveal the intensity of students' affection or disaffection toward mathematics.

Metaphor Assessment Instrument

Figure 10.1 shows the metaphor survey instrument, which asked each student to show his or her dispositions toward mathematics by using food, colors, and animals as metaphors. The tool instructed students to take time to think about the questions and describe how they truly felt about mathematics. Most important, students had to describe the reasons why they chose specific food, colors, and animals. The advantage of asking students to supply three metaphors is that it allows teachers to better gauge the breadth and depth of students' dispositions toward mathematics. Since younger students tend to think literally, using metaphor assessments is more appropriate for middle and high school students because of their maturity in writing and ability to think metaphorically.

We are interested in learning how you think and feel about mathematics. Please take a few minutes to think about the following questions and write how you truly feel. There are no right or wrong answers.

If math were a food, it would be ______________________________
because ____
____
____

If math were a color, it would be ______________________________
because ____
____
____

If math were an animal, it would be ___________________________
because ____
____
____

Fig. 10.1. Metaphor survey instrument

In our research, we administered the instrument to all 1300 students from Lennon High School, an ethnically diverse inner-ring suburban school located in the greater Philadelphia area. Its student population is approximately 40 percent African American, 30 percent Hispanic, 25 percent white, and 5 percent Asian. Less than 20 percent of the community has a college degree, and more than 40 percent of the school's students are considered economically disadvantaged. Overall, about 55 percent of the eleventh-grade students in the school score proficient or advanced on the state mathematics test. Students were in three tracks: (1) college-preparatory track using a Standards-based, National Science Foundation–funded curriculum; (2) college-preparatory track using a non–Standards-based curriculum; and (3) non–college-preparatory track using a non–Standards-based curriculum. The Standards-based curriculum was developed with the support of the National Science Foundation and implemented to align with the recommendations of the NCTM Standards.

Two Methods for Analyzing Metaphors

We analyzed the responses to the metaphor survey instrument both qualitatively and quantitatively. For the qualitative analysis, we categorized each student's metaphor as food, colors, and animals to determine what kinds of metaphors students used and why. The qualitative analysis allowed for an in-depth understanding of students' dispositions toward mathematics as well as *why* they developed such dispositions. For the quantitative analysis, we developed a holistic scoring rubric (table 10.1) to score students' responses with a five-point scale (1–5).

Table 10.1
Holistic scoring rubric for metaphors

Score level (points)	Description	Example
1	Very negative	"Like rotten cheese. It is disgusting and I hate it."
2	Moderately negative	"Brown is my least favorite color, and math is my least favorite class."
3	Neutral or ambivalent	"It's okay. I don't like it but I know I need it."
4	Moderately positive	"I like math like I do pizza."
5	Very positive	"My birthstone, purple, is my favorite color. It brings passion. That's how I feel about math."

We scored students' levels of dispositions toward mathematics. A pair of mathematics teachers rated each student's response. Interrater reliability was more than 95 percent. The quantitative analysis yields an overall picture about a group of students' dispositions toward mathematics and simplifies statistical analysis.

Students displayed a remarkably similar level of dispositions toward mathematics across all three metaphor categories. No student showed opposite dispositions toward mathematics by using different metaphors; that is, students did not mix a sunny color metaphor with a rotten food metaphor.

Figure 10.2 shows the percentage distributions of students in each score level. Students who were enrolled in different courses and levels demonstrated different mathematical dispositions with metaphors. Students in the college-preparatory track using a Standards-based curriculum had more positive dispositions toward mathematics than those in non–college-preparatory track and college-preparatory track using a non–Standards-based curriculum. Although roughly 10 percent of each group of students reported "very positive" dispositions toward mathematics (level 5), students in the college-preparatory track with the Standards-based curriculum were less likely than students in the other groups to report "very negative" or "moderately negative" attitudes (level 1 or 2) and more likely to report "neutral" or "moderately positive" attitudes (level 3 or 4).

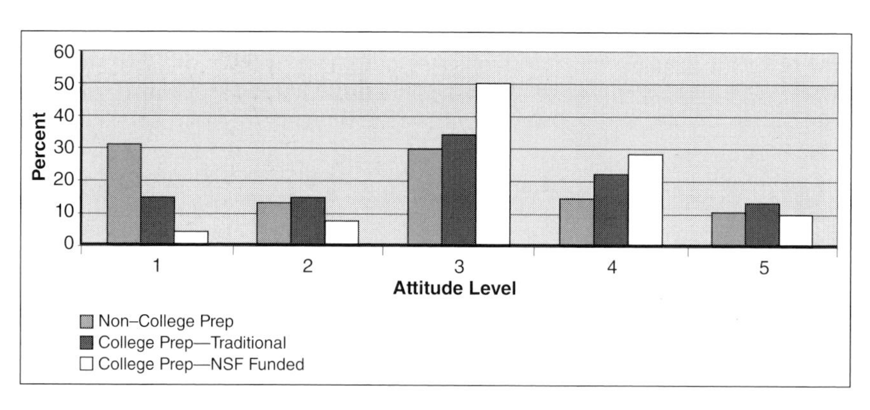

Fig. 10.2. Percent distribution at each attitude level

Kinds of Metaphors and Mathematical Dispositions

The depth and passions that many students' metaphors expressed surprised us. Students used a variety of food, color, and animal imagery to express how they felt about mathematics. However, their explanations of *why* they chose a particular metaphor revealed their inner dispositions toward mathematics. Our qualitative analysis suggests that the responses for the food and animal metaphors are more robust than those for color metaphors. Sometimes students did not specifically use a food, color, or animal; instead, they used terms such as "tasty food," "sad color," or "pets" to describe their feelings about mathematics. The following sections offer examples and analyses of students' metaphors.

To Show *How* They Like Mathematics

Students used metaphors in often-striking ways to express their love or hate about math. Following are examples of food, color, or animal metaphors showing how students expressed their positive dispositions toward mathematics:

- "[Math] is like [Nachos Bell Grande], because I love a [Nachos Bell Grande] from Taco Bell. It [is] always the same, but at the same time it can surprise you with a new taste."
- "Cheese, yummy! I have cheese . . . every day."
- "Purple is my favorite color. It's my birthstone color; plus, it brings [passion]. That's how I feel about math."
- "Red is my favorite color, and I like to wear and color with red. I like math, so if math could be a color, it [would] be red."
- "Dogs! Because [math] acts as your companion, assisting you with everyday difficulties without you expecting it. Math can be an enjoyable subject if you put the time and effort into it. It takes hard work to develop, but the ending result is so rewarding."
- "I would say monkeys. Because monkeys are very intelligent animals, and if they learn something, they do it [well]. Well math is needed, you learn it, you have to be dedicated to [it]."

To Show *Why* They Like Mathematics

Students gave two major reasons why they like mathematics. First, students liked or enjoyed mathematics because they experienced the satisfaction of overcoming the challenges and struggles to achieve their goals:

- "Math is like steak because math is a full, expansive subject. However, like a steak there are tough bits of gristle scattered throughout—obstacles you must work around. The full meal is satisfying, but the process of eating is somewhat unusually strenuous."
- "Pineapple has a rough outer shell that is hard to break, but once the shell is broken there is a great satisfaction. Math is the same. It is tough to start, but once you got it, it is fun."

- "If math were an animal, it would be a lion because it can be scary at first, but it can eventually be tamed and [you can] overcome the fear."
- "Like a dog takes time to train. Math takes time to learn, and it may be difficult to understand at times, but it helps."

The second reason for students to like or enjoy mathematics is utilitarian: mathematics is important in daily life. Many students used their metaphors to explicitly state the importance of mathematics in their daily life and society:

- "I enjoy mathematics. I like the challenge it brings. Most importantly though, I believe math is the foundation of many aspects in life. You cannot go a single day without encountering math. I intend on using math in my future career. I think of math like a puzzle—there are many different pieces that fit together nicely to make a whole. The different subjects in mathematics not only [complement] each other but other subjects in school as well."
- "[An] apple is healthy, and math is healthy for the mind."
- "Vegetables are good for you, and so is mathematics for daily things. It is needed in life. Some people like it, and some people don't, but you still need it to live a healthy life."
- "If math were a food, it would be water because we need mathematics in life like [we need] water. To me, math is my favorite subject."
- "[A] lion has power. Through practice in its studies, one can emerge a strong wielder of its powerful knowledge—same for math."

To Show *How* They Disliked Mathematics

Similarly, students used metaphors to clearly express their intense dislike for mathematics:

- "Like rotten cheese, it is disgusting. Math is not that much fun, and it would not be that tasty because nobody would like it."
- "Black is a dark color that not a lot of us can describe or understand. Math to me is a dark shadow that I don't like."
- "I would say a mosquito, because whatever you do to try and get away from it, it always comes back. It's annoying because you hate taking math every year, and whatever you try to do to stop it, it always fails."

To Show *Why* They Dislike Mathematics

From the analysis of the metaphors, we found two reasons why students dislike mathematics: (1) they cannot overcome the challenges they face in the learning of mathematics; (2) they do not see the usefulness of mathematics.

- "Tiger. Because I don't like tigers: ferocious and [easy] to fear—just like math with too much thinking and hard-to-solve problems. It really hurts my brains to think about all those problems that I cannot solve."
- "It is like gum. You chew gum and use it to freshen up your breath, but in the end, it's worthless and doesn't have any nutrition or vitamins. Math is used in school to determine your intelligence, but there is no need for it later."

To Show Their Views about the Nature of Mathematics

In addition to showing whether and why students like or dislike mathematics, metaphors also show students' views about the nature of mathematics. In their explanations, students usually tried to connect their metaphors with a certain aspect of mathematics. For example, students used toppings of a pizza to show many different components in mathematics:

- "Just as a pizza is filled with plenty of different vegetables, math [consists] of several different equations and functions that go together to create a well-rounded subject."
- "The color white is all the colors blended into one. In math, you integrate everything that you've learned in the past with the new things that you learn now."
- "Because yellow has an interesting significance—meaning both honor and cowardice. Math, too, is two-faced. It represents the natural world, but it is also remarkably manmade."

Classroom Tips for Using Metaphors to Assess Mathematical Dispositions

For convenience of analysis, giving individual students a set of metaphor types is preferable to asking them to invent their own categories. If you do not use food, color, and animal categories, choose metaphor categories that students are familiar with and can relate to. Other metaphors to use could be sports, cars, games, and movies. In different grade levels, teachers may use different metaphors. The set of chosen metaphors must give students the opportunity to use rich "vocabularies" to describe their dispositions toward mathematics.

When asking students to describe mathematics by using a set of metaphorical categories, having students state why they chose to use a particular metaphor is important. Two students might use the same metaphor for different reasons.

For example, many students may say that "mathematics is like pizza." But one group may say "pizza" because it is the students' favorite food, whereas others may say "pizza" because it is their least favorite food.

When analyzing their students' responses, teachers may discover patterns by grade level, type of mathematics course, tracking level, gender, race/ethnicity, neighborhood, and student peer group. For example, as experienced teachers know, each classroom develops its own personality and student culture toward schooling in general and mathematics learning in particular. Particular classrooms will probably have several student cliques made up of students who share similar dispositions toward mathematics that are mutually reinforcing, either positive or negative. A metaphor survey can help reveal the sociology within a classroom.

Gauging students' dispositions toward learning a particular subject should be an integral part of classroom instruction, especially in the beginning of the academic year when a classroom culture is forming. Teachers may elect to discuss the survey results of mathematical dispositions with their class as a way to uncover tendencies to resist learning. Individual or group discussion can reinforce positive mathematical dispositions and help defuse negative ones by merely acknowledging them. In this article, we introduced two methods to analyze students' metaphors. These two methods are related and complementary to facilitate our understanding of students' mathematical dispositions. Teachers may also ask students to score their own metaphors (self-scoring) or score metaphors for each other (peer scoring). Both approaches give students another opportunity to understand and make public their level of mathematical dispositions.

Conclusion

This paper described how using metaphors can be a powerful means to assess students' dispositions toward mathematics. In particular, we discussed using familiar items as metaphors: foods, colors, and animals. Using metaphors offers students a much richer context and vocabulary to express and communicate the nature and intensity of their dispositions toward mathematics than do multiple-choice surveys. By using metaphor surveys, teachers can develop better strategies to engage reluctant learners and win over complacent student peer cultures.

We also introduced two methods to analyze students' metaphors. The qualitative analysis helps us understand not only whether but also why students like or dislike mathematics. The quantitative analysis helps us understand the intensity of students' dispositions. Our analysis of metaphors showed that if students made the effort and overcame the struggle in learning mathematics, they tended to enjoy doing mathematics and solving problems. The sense of satisfaction with success motivated students to love mathematics and continue to try to overcome challenges. This dynamic leads to a positive disposition–performance cycle: a *virtuous* cycle. Experiencing and overcoming challenges gives students a sense of accomplishment, which in turn motivates them to accept new challenges with enthusiasm.

Whether students can see the value of mathematics—and connect what they are learning with the needs of their personal lives and future careers—also affects their dispositions. The missed opportunities of seeing the usefulness or importance of mathematics learned in school could lead to more negative dispositions toward mathematics. Under such circumstances, students would view mathematics as a pointless exercise that they would not need later in life. Thus, to improve students' mathematical disposition, teachers must help students experience the success of solving mathematical problems and see the importance of learning mathematics.

(We thank Eileen M. Egan for assistance with data collection. The National Science Foundation [grant no. ESI-0314806] supported this research. Any opinions are the authors' and do not necessarily represent the views of the National Science Foundation.)

REFERENCES

Bullough, Robert V. "Exploring Personal Teaching Metaphors in Preservice Teacher Education." *Journal of Teacher Education* 42 (January–February 1991): 43–51.

Cooney, Thomas J., Fred Goffrey, Max Stephens, and Marilyn Nickson. "The Professional Life of Teachers." *For the Learning of Mathematics* 5 (April 1985): 24–30.

English, Lyn D., ed. *Mathematical Reasoning: Analogies, Metaphors, and Images*. Mahwah, N.J.: Erlbaum, 1997.

Lakoff, George, and Mark Johnson. *Metaphors We Live By*. Chicago: University of Chicago Press, 1980.

Merlino, F. Joseph. "Understanding Integrated Mathematics Using Living Metaphors." *NCTM Mathematics Education Dialogues*. January 2001. www.nctm.org/resources/content.aspx?id=1674 (accessed June 10, 2010).

Miller, Steven I., and Marcel Fredericks. "Uses of Metaphor: A Qualitative Case Study." *Qualitative Study in Education* 1 (July 1988): 263–76.

Munby, Hugh. "Metaphor in the Thinking of Teachers: An Exploratory Study." *Journal of Curriculum Studies* 18 (May 1986): 197–209.

National Council of Teachers of Mathematics (NCTM). *Curriculum and Evaluation Standards for School Mathematics*. Reston, Va.: NCTM, 1989.

Presmeg, Norma C. "Prototypes, Metaphors, Metonymies, and Imaginative Rationality in High School Mathematics." *Educational Studies in Mathematics* 23 (December 1992): 595–610.

Sfard, Anna. "On Two Metaphors for Learning and the Dangers of Choosing Just One." *Educational Researcher* 27 (March 1998): 4–13.

Wolodko, Brenda L., Katherine J. Willson, and Richard E. Johnson. "Metaphors as a Vehicle for Exploring Preservice Teachers' Perceptions of Mathematics." *Teaching Children Mathematics* 10 (December 2003): 224–29.

Chapter 11

Using Prediction to Motivate Personal Investment in Problem Solving

Lisa Kasmer
Ok-Kyeong Kim

NEARLY ALL educators agree that motivation significantly influences student engagement in learning and later learning outcomes (Haselhuhn, Al-Mabuk, and Gabriele 2007). Paradoxically, motivation declines as students transition from elementary to middle school (Eccles and Midgley 1989; Haselhuhn, Al-Mabuk, and Gabriele 2007). This effect becomes problematic in middle school mathematics if students are not motivated to learn while wrestling with the increased cognitive demands of the mathematics at this level. In this chapter, we highlight an instructional strategy in middle school mathematics classrooms to demonstrate how teachers can enhance motivation. However, we believe that our example can work for other grade levels.

Mastery Orientation versus Performance Orientation

According to Anderman and Midgley (1998), students enter academic pursuits primarily for two reasons, "mastery learning or task goal orientation, and performance or ability goal orientation" (p. 10). Ames (1992) suggested that mastery orientation and performance orientation offer divergent perspectives of success and rationale for entering a task. Students whose motivation is to develop an understanding and competence exemplify mastery orientation. This intrinsic motivation to learn orients students to continually strive for understanding and development of new skills (Ames 1992). In these situations, students become personally invested in their own learning. Ames further acknowledged, "research evidence suggests that a mastery goal is associated with a wide range of motivation-related variables" (p. 262) such as expending effort, which leads to success and positive attitudes about learning (Ames and Archer 1988). Furthermore, a mastery goal orientation, as Butler (1987) suggested, increases the time that students will devote to learning tasks, such as mathematical explorations. An increased motivation to learn and understand new material drives this willingness to dedicate more time. These motivation-related variables can contribute to achievement by fostering tenacity and perseverance to tackle more challenging mathematics.

In contrast, performance-oriented students approach tasks to prove competence or to prevent seeming incompetent to their peers (Barber and Olsen 2004; Dweck 1986). Covington (1984) suggested that performance-oriented students regard public recognition of outperforming others to be the core of their success. Nolen (1988) believed that performance orientation is associated with learning behaviors such as memorization of procedures and rote strategies rather than learning for understanding. Because such students do not necessarily value understanding of the content, they do not personally invest.

Mastery-oriented students regard errors as essential to learning (Meyer, Turner, and Spencer 1997). Consider the student who makes errors on a mathematics assessment. A mastery-orientation student does not associate mistakes with failure, whereas a performance-oriented student views mistakes as failure or incompetence (Maeher and Midgley 1996). Students with a mastery-orientation perspective tend to be intrinsically motivated or more willing to engage in challenging tasks and display more positive feelings about themselves as learners (Haselhuhn, Al-Mabuk, and Gabriele 2007). Teachers' carrying out purposeful instructional moves in their classrooms can cultivate this personal investment.

Wenglinsky (2002) indicated that teaching practices seemed to have more influence on student learning than socioeconomic status, at least as the National Assessment of Educational Progress measured student outcomes. Teacher instructional practices can also influence students' goal orientations (Anderman, Patrick, and Ryan 2004; Patrick et al. 2001). These goal orientations in turn affect a student's motivation to learn. When students face challenging tasks that attend to understanding mathematics (rather than procedural memorization) and require making sense of mathematics, the goal structure shifts toward a mastery orientation. Once this goal structure is in place, students may become intrinsically motivated to learn.

Student Motivation and Using Prediction

Teachers can support mastery orientation by giving students opportunities to be active learners who regulate their understanding, connect new learning to prior knowledge, and can self-correct (Anderman, Patrick, and Ryan 2004). Anderman and Anderman (2010) suggest that these "deep cognitive strategies" help promote motivation. To support opportunities for active learning and foster connections that inspire mastery orientation and personal investment, teachers can encourage students to predict results and discuss these predictions along with supportive reasoning. Students have an opportunity to self-correct as they reconcile any discrepancy between their initial predictions and the results of an exploration. Incorporating prediction prompts in mathematics lessons helps spark students' personal investment in the problem, which will inherently increase motivation. Prediction can increase students' personal investment because "the commitment involved in deciding on a prediction can have powerful motivation effects" (White and Gunstone 1992, p. 63). Garfield and Ben-Zvi (2007) illustrate this notion of personal investment in their suggestion that "if students are first asked to make guesses or predictions about data and random events, they are more likely to care about and process the actual results" (p. 388).

Various disciplines have validated prediction. In science education, Lavoie (1999) studied the effects of supplementing lessons with a *prediction/discussion phase* before the three-phase learning cycle (exploration, term introduction, and concept application). Including *hypothetico-predictive* reasoning "produced significant gains relative to the use of process skills, logical-thinking skills, science concepts, and scientific attitudes" (p. 1127). For reading, Duffy (2003) notes that reading comprehension involves an active cycle that begins with prediction. He found that readers anticipate meaning by making predictions and refining their predictions during their reading, as well as that asking students to predict the outcome of a story gives students a sense of engagement and purpose for reading. Thomas-Fair (2005) also found that prediction benefits readers as early as kindergarten. She noted that posing prediction questions offered opportunities to activate prior knowledge, motivate students, and encourage their natural curiosity.

In mathematics education, we define prediction as previewing the mathematical ideas of a lesson and using prior knowledge to consider what results are reasonable. For example, before students explore a problem involving the relationship between distance and time, teachers may ask them to predict how increasing walking rates will affect the table, the graph, and the equation. Such questions could help motivate the students to begin the problem and seek a solution. Because students view prediction questions as not having a predetermined correct response, such questions can also afford students an opportunity to engage in classroom discourse without fear of being incorrect. Eliminating the pressure of producing the correct response can help students' goal orientation shift toward mastery orientation rather than performance orientation. Students become personally invested in the problem as they begin to gauge the reasonableness of their own predictions.

A Classroom Example

Consider a classroom vignette (Kim and Kasmer 2007a) in which middle school students solved a problem (adapted from Lappan et al. 1998) involving a race between two brothers: "Emile's walking rate is 2.5 meters per second, and his little brother Henri's walking rate is 1 meter per second. Henri challenges Emile to a walking race. Emile gives Henri a 45-meter head start. How long should the race be so that Henri will win by just a bit?"

Before the students begin to solve the problem, the teacher asked them to make various predictions related to this problem, write down their predictions and supportive reasoning, and then discuss the reasonableness of each prediction. First, they predicted whether graphs, tables, and equations for this problem would look similar to what they had done before. One student said that this problem would produce lines with constant rates, which made them "linear." Other students agreed and said that some problems they had done involved constant rates and some did not. Next, they predicted which line would be steeper if they graphed the situation. One student said, "Henri's got steeper because he has a 45-meter head start." Another student offered, "I think Emile because he goes farther and faster in a shorter time." When the teacher asked how she knew he went farther and faster in less time, the student answered, "Because he's a lot faster." Some students agreed. One student said, "Because he goes 2.5 meters per second and he travels faster, so his line will be steeper." Last, students predicted how long they thought the race should be. Students offered various predictions ranging from 50 to 250 meters. One student, whose prediction was 250 meters, said, "There should be a longer distance so Emile can catch up." As soon as he finished, another student said, "I disagree with him because Emile's walking rate is double Henri's. So, it's not going to be 100 and up." Many agreed and said, "100 is too high" and changed their predictions for the race distance. A couple of students were trying to figure out the distance that each could make in a certain time (i.e., 10 seconds). Then the teacher asked pairs of students to figure out the problem by using the ideas that they had discussed. In this class, every student pair began the problem immediately, the teacher having surmised that each had a viable entry ramp to the problem and were confident and motivated to begin.

In this particular classroom example, constructing predictions before exploring the problem guided students toward a mastery orientation. The prediction prompt contributed to the students' focus on the mathematical meaning and structure of the problem, thus leading students to an authentic mathematical exploration rather than a skill-oriented routine. This approach motivated students not only to find what the answer to the problem would be but also to discuss and reason about what mathematical ideas this problem encompassed. A nonevaluative stance toward student predictions can motivate all students to engage in and become personally invested in the mathematics of the problem. Ames concurs that "the ways in which students are evaluated is one of the most salient classroom factors that can affect student motivation" (1992, p. 264).

The prediction opportunity in this example also explicitly helped the students build links between previous knowledge (i.e., what they knew about linear relationships—constant rates of change) and a new concept (i.e., another aspect of a linear function—y-intercept represented as a 45-meter head start). They were also establishing connections among the mathematical ideas, the problem context, and various representations of the ideas (e.g., what it means to have two different rates of change and how their graphs look different). Making such connections is one underlying structure of mastery orientation (Anderman, Patrick, and Ryan 2004).

Teacher Perspective on Prediction's Impact on Motivation

In this section, we highlight some significant teacher input relative to students' motivation from Kasmer's (2008) study that compared a class that incorporated prediction into lessons with one that did not. The same teacher taught both classes with the same curriculum. Classroom videotapes validated the teacher's claims in interviews and journal entries. Analysis revealed that the students in the prediction classroom seemed more motivated and engaged, because they were more confident and eager to enter the problem, than students in the nonprediction classroom. The students in the prediction class, while not knowing the solution to the problem before the exploration component of the lesson, were willing to begin this segment of the lesson with a preliminary conception of how to proceed. They had motivation and desire to determine whether their prediction responses were reasonable. Students understood that mistakes were inherent in their prediction responses and that the summary component of the lesson would address resolving these mistakes.

The teacher's comments throughout the study revealed a notable distinction between the prediction and nonprediction classes relative to student motivation. She attributed this distinction to the expectation that each student would have a well-thought-out response to the prediction question(s) that she posed. Furthermore, students appeared to feel more confident responding. She believed that this confidence was due to students' having had an opportunity to organize their thinking before responding to the prediction prompt. This organizing, in turn, gave them a chance to begin to consider the mathematics of the lesson. The teacher also observed that the students in the prediction class generally responded with more depth and detail than the nonprediction class students: the prediction responses seemed to foster a personal investment in the problem, and students wanted to share their thinking.

Suggestions

The classroom vignette illustrates how prediction can guide students toward mastery orientation, thus capitalizing on student intrinsic motivation. In preparing a lesson, teachers should be keenly aware of opportunities to incorporate prediction questions to motivate students to engage in the mathematics of the lesson. However, teachers must not only furnish prediction questions at the onset of the lesson but also revisit the students' prediction responses and reconcile discrepancies after exploring the problem. For this strategy of prediction to be effective, teachers should consider the following conditions.

Create an Inviting Classroom Culture

At the beginning of the school year, teachers need to create a classroom environment where students feel comfortable taking risks and making conjectures. Teachers should establish norms of interaction to reassure students that all prediction responses have value and that the process requires supportive reasoning. Prediction responses should be plausible ideas as opposed to mere guesses without appropriate support. To move past the notion of seeking only a correct response—and to shift toward a focus on developing understanding—students need encouragement to share ideas with one another and constructively evaluate each other's ideas.

Deliberately Plan to Incorporate Prediction Questions into Lessons

When deciding to use prediction questions, teachers should consider the content of the lesson. Prediction questions should implicitly reflect the mathematical content of problems posed without revealing the essence of those problems. Table 11.1 shows sample prediction questions along with contexts. The teacher then elicits student responses without commenting on the accuracy of the prediction or the appropriateness of reasoning. Teachers should encourage students to comment on others' ideas. The lesson's summary segment should revisit the prediction questions and student responses. Students should then reconcile any discrepancies between their initial predictions and the outcome of the problem.

Have Students Write Their Responses to Prediction Prompts before Class Discussion

Individual written responses are desirable as evidence of student thinking, since time constraints do not allow students to share their predictions during the launch of the lesson. Furthermore, writing individual responses allows students to organize their thinking about the mathematics of the problem before verbalizing their

Table 11.1
Sample prediction questions and context

Prediction question	Context
Which number has more factors, 24 or 50? Provide your reasoning.	Students explore factors of numbers.
Can you predict the impact of b on the appearance of the line $y = ax + b$? What did you think about to help you make this prediction?	Students explore the y-intercept with graphing calculators in the context of linear relationships.
Do you predict that this situation will be linear or exponential? How do you know? What did you think about to help you make this prediction?	Students explore exponential relationships that involve examining repeated doubling and tripling.
Can you predict what the graph of this situation might look like? Can you predict the maximum area of the shape? What is your prediction based on?	Students explore a fixed perimeter and the relationship between length and area.
What is the result of putting together two-dimensional and three-dimensional shapes? Provide your reasoning.	Students consider the results of combining two-dimensional and three-dimensional figures.
Predict whether the mean is to the right or left of the median. Why is this happening? Predict what happens to the mean and the median if you remove the outlier.	Students explore the mean and the median of a data set represented on a line plot.

thinking to the entire class. Here the focus is on understanding the mathematics (i.e., mastery orientation). Requiring students to respond in writing to the prediction questions helps students use their own reasoning rather than waiting for those of classmates.

Use Prediction in a Variety of Mathematical Situations

Prediction questions can motivate students in all content strands of mathematics. The classroom vignette used algebra. Data analysis and probability can also easily incorporate prediction questions and motivate students to engage in and think about the mathematics (Kim and Kasmer 2007b; Jones et al. 1999). For instance, an e-example from *Principles and Standards for School Mathematics* (National Council of Teachers of Mathematics [NCTM] 2008a) shows that students can predict and justify what will happen to the mean and the median of the given data when one data value is pulled toward one end or the other end (fig. 11.1). Students can also predict the data set that would produce a mean that does not represent the data.

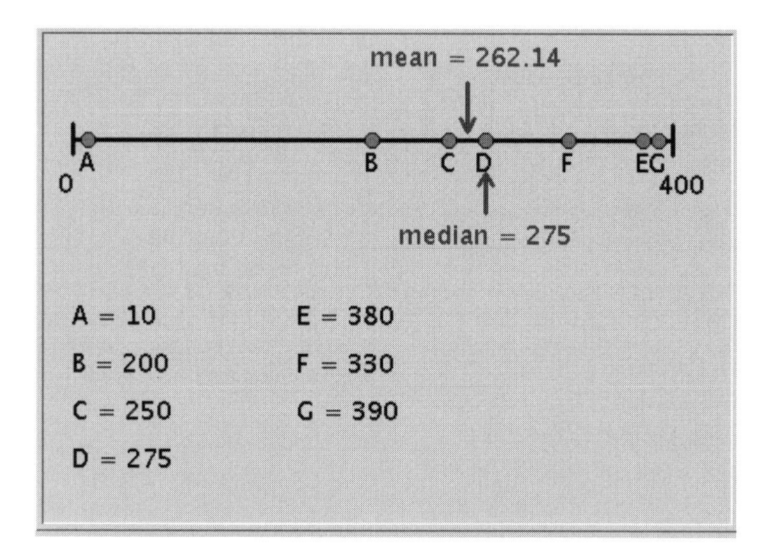

Fig. 11.1. Relating mean and median

Many geometry problems also lead to rich prediction questions. For example, a teacher can ask students to predict what solid they generate from a given net pattern (fig. 11.2). Here the teacher presents the net to students and asks them to visualize the solid and justify their predictions. The teacher might also ask students to predict whether other nets that would produce the same solid exist. Teachers can extend this activity by using various solids and corresponding flat patterns. Once students have made and discussed their predictions, students would receive flat patterns to construct their own solids.

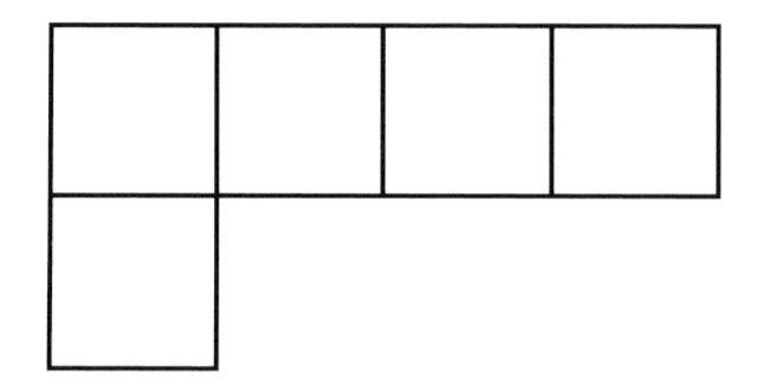

Fig. 11.2. Net pattern

In another geometry task adapted from NCTM's Illuminations (NCTM 2008b), students work in groups to place the minimum number of security cameras needed in each room (i.e., variously shaped polygons) to monitor all parts of the room. Students must make the connection between the number of vertices in a polygon and the number of cameras needed to monitor the area of the room. Finally, from this investigation students develop the formula for the maximum number of cameras needed for an n-gon. In addition to the questions that the Illuminations Web site offers, a teacher may also consider requests such as "Predict whether all six-sided rooms require the same or a different minimum number of security cameras. Explain your reasoning."

Conclusion

When students are motivated intrinsically and personally invested in mathematical tasks, they exhibit more pedagogically desirable behaviors (Middleton and Spanias 1999). Asking students to make predictions and discussing related ideas provokes and increases students' learning opportunities by enhancing engagement and motivation. When classroom teachers use prediction as an instructional strategy, they create a learning environment where students can access prior knowledge, connect mathematical ideas, and actively and confidently engage in problem solving and discussion. Students also begin to value learning, not just getting correct responses. As a result, giving students such opportunities helps shift goal orientation toward mastery rather than performance. When mastery learning is the fundamental goal of learning, students can and will be motivated to undertake and personally invest in the challenges presented.

REFERENCES

Ames, Carole. "Classrooms: Goals, Structures, and Student Motivation." *Journal of Educational Psychology* 84 (September 1992): 261–71.

Ames, Carole, and Jennifer Archer. "Achievement Goals in the Classroom: Students' Learning Strategies and Motivation Processes." *Journal of Education Psychology* 80 (September 1988): 260–67.

Anderman, Eric, and Lynley Anderman. *Classroom Motivation*. Upper Saddle River, N.J.: Pearson, 2010.

Anderman, Eric, and Carol Midgley. "Motivation and Middle School Students." Champaign, Ill.: ERIC Clearinghouse on Elementary and Early Childhood Education, 1998. ERIC Document Reproduction Service no. ED421281.

Anderman, Lynley H., Helen Patrick, and Allison M. Ryan. "Creating Adaptive Motivational Environments in the Middle Grades." *Middle School Journal* 35 (May 2004): 33–39.

Barber, Brian K., and Joseph A. Olsen. "Assessing the Transitions to Middle School and High School." *Journal of Adolescent Research* 19 (November 2004): 3–30.

Butler, Ruth. "Task-Involving and Ego-Involving Properties of Evaluation: Effects of Different Feedback Conditions on Motivational Perceptions, Interest, and Performance." *Journal of Educational Psychology* 79 (December 1987): 474–82.

Covington, Martin V. "The Self-Worth Theory of Achievement Motivation: Findings and Implications." *Elementary School Journal 85* (September 1984): 5–20.

Duffy, Gerald E. *Explaining Reading: A Resource for Teaching Concepts, Skills, and Strategies*. New York: Guilford Press, 2003.

Dweck, Carol S. "Motivational Processes Affecting Learning." *American Psychologist* 41 (October 1986): 1040–48.

Eccles, Jacquelynne S., and Carol Midgley. "Stage-Environment Fit: Developmentally Appropriate Classrooms for Young Adolescents." In *Research on Motivation in Education: Goals and Cognitions*, edited by Carole Ames and Russell Ames, pp. 139–86. New York: Academic Press, 1989.

Garfield, J., and D. Ben-Zvi. "The Discipline of Statistics Education." In *Background Papers of the Joint ICMI/IASE Study on Statistics Education in School Mathematics: Challenges for Teaching and Teacher Education*, edited by C. Batanero. Granada, Spain: University of Granada, 2007.

Haselhuhn, Charlotte, Radhi Al-Mabuk, and Anthony Gabriele. "Promoting Positive Achievement in the Middle School: A Look at Teachers' Motivational Knowledge, Beliefs, and Teaching Practices." *Research in Middle Level Education* 30, no. 9 (2007): 1–20.

Jones, Graham A., Carol A. Thornton, Cynthia Langrall, and James Tarr. "Understanding Students' Probabilistic Reasoning." In *Developing Mathematical Reasoning in Grades K–12*, 1999 Yearbook of the National Council of Teachers of Mathematics (NCTM), edited by Lee V. Stiff and Frances R. Curcio, pp. 146–55. Reston, Va.: NCTM, 1999.

Kasmer, Lisa. "The Role of Prediction in the Teaching and Learning of Algebra." Ph.D. diss., Western Michigan University, 2008.

Kim, Ok-Kyeong, and Lisa Kasmer. "Prediction and Mathematical Reasoning." In *Proceedings of the 5th Hawaii International Conference on Education* (2007a). www.hiceducation.org/EDU2007.pdf (accessed December 17, 2010).

———. "Using Prediction to Promote Mathematical Reasoning." *Mathematics Teaching in the Middle School* 12 (February 2007b): 294–99.

Lappan, Glenda, Jim Fey, William Fitzgerald, Susan Friel, and Elizabeth D. Phillips. *Connected Mathematics*. Needham, Mass.: Pearson Prentice Hall, 1998.

Lavoie, Derrick. "Effects of Emphasizing Hypothetico-Predictive Reasoning within the Science Learning Cycle on High School Students' Process Skills and Conceptual Understandings in Biology." *Journal of Research in Science Teaching* 36 (December 1999): 1127–47.

Maeher, Martin L., and Carol Midgley. *Transforming School Cultures*. Boulder, Colo.: Westview Press, 1996.

Meyer, Debra K., Julianne C. Turner, and Cynthia A. Spencer. "Challenge in a Mathematics Classroom: Students' Motivation and Strategies in Project-Based Learning." *Elementary School Journal* 97 (May 1997): 501–22.

Middleton, James, and Photinia Spanias. "Motivation for Achievement in Mathematics: Findings, Generalizations, and Criticisms of the Research." *Journal for Research in Mathematics Education* 30 (January 1999): 65–88.

National Council of Teachers of Mathematics. "E-Example 6.6: Comparing Properties of the Mean and the Median with Technology." 2008a. www.nctm.org /fullstandards/document/eexamples/chap6/6.6/default.asp (accessed October 20, 2008).

———. "Illuminations: Security Cameras." 2008b. illuminations.nctm.org /LessonDetail.aspx?id=L767 (accessed November 4, 2008).

Nolen, Susan B. "Reasons for Studying: Motivational Orientations and Study Strategies." *Cognition and Instruction* 5 (December 1988): 269–87.

Patrick, Helen, Lynley H. Anderman, Allison M. Ryan, Kimberly C. Edelin, and Carol Midgley. "Teachers' Communication of Goal Orientations in Four Fifth-Grade Classrooms." *Elementary School Journal* 102 (September 2001): 35–58.

Thomas-Fair, U. "The Power of Prediction: Using Prediction Journals to Increase Comprehension in Kindergarten." Paper presented at the Georgia Association of Young Children Conference, Atlanta, 2005.

Wenglinsky, Howard. "How Schools Matter: The Link between Teacher Classroom Practices and Student Academic Performance." *Educational Policy Analysis Archives*. 2002. epaa.asu.edu/ojs/article/view/291 (accessed June 10, 2010).

White, Richard T, and Richard F. Gunstone. *Probing Understanding.* London: Falmer Press, 1992.

Chapter 12

Developing Persistent and Flexible Problem Solvers with a Growth Mindset

Jennifer Suh
Stacy Graham
Terry Ferrarone
Gwen Kopeinig
Brooke Bertholet

I just kept going like a snow plow stuck in the road. I didn't wait for the spring to come. I kept going.

—Griffin, fifth grade

THIS CHAPTER describes research from a group of mathematics teachers and a university researcher who collaborated through Lesson Study, a form of professional development that focuses on research lessons. At the beginning of our Lesson Study, we developed our research aim and overarching goal: to develop persistent and flexible problem solvers. Through collaborative planning and designing of problem-driven lessons throughout the academic year, the teacher–researchers developed classroom communities of inquiry and specific strategies that promoted students' persistence and flexible thinking in problem solving. Teachers observed marked progression in students' productive dispositions toward mathematics throughout the school year. Students developed a "growth mindset" (Dweck 2006) focused on effort and persistence in learning mathematics. The Lesson Study model of professional development also influenced teachers' instructional practices in developing persistent and flexible problem solvers.

Research Background

According to Dweck (2006), a Stanford University psychologist with three decades of research on achievement and success, two mindsets about learning exist: a *growth mindset* and a *fixed mindset*. A growth mindset holds that your basic qualities are things that you can cultivate through your efforts, whereas a fixed mindset holds that your qualities are "carved in stone" (i.e., your intelligence is something you can't change very much). When facing challenging problems, children who believe that effort drives intelligence tend to do better than children who believe that intelligence is a fixed quality that they cannot change. According to research on competence and motivation (Elliot and Dweck 2005; Weiner 2005), students can attribute their successes and failures to ability (e.g., "I'm just [good/bad] at mathematics"), effort (e.g., "I [worked/did not work] hard enough"), luck, or powerful people (e.g., "the teacher [loves/hates] me"). A student with a fixed mindset avoids challenges, gives up easily, sees effort as fruitless or worse, ignores useful negative feedback, and feels threatened by the success of others. Meanwhile, a student with a growth mindset embraces challenge, persists despite setbacks, sees effort as the path to mastery, learns from mistakes and criticisms, and finds lessons and inspiration in the success of others. People with a growth mindset believe that they can develop their abilities through hard work, persistence, and dedication; brains and talents are merely a starting base (Dweck 2006).

Research also suggests that good problem solvers are qualitatively different from poor problem solvers (National Research Council 2004; Schoenfeld 2007). Good problem solvers are flexible and resourceful. They have many ways to think about problems: "alternative approaches if they get stuck, ways of making progress when they hit roadblocks, of being efficient with (and making use of) what they know. They also have a certain kind of mathematical disposition—a willingness to pit themselves against difficult mathematical challenges under the assumption that they will be able to make progress on them, and the tenacity to keep at the task when others have given up" (Schoenfeld 2007, p. 60). Problem solvers experience a range of emotions associated with different stages in the solution process. Mathematicians who successfully solve problems say that having done so contributes to an appreciation for the "power and beauty of mathematics" (National Council of Teachers of Mathematics [NCTM] 1989, p. 77) and the "joy of banging your head against a mathematical wall, and then discovering that there might be ways of either going around or over that wall" (Olkin and Schoenfeld 1994, p. 43). Good problem solvers also are more willing to engage with a task for a length of time, so that the task ceases to be a "puzzle" and becomes a problem (Schoenfeld 2007).

Creating opportunities for success in mathematics is important, but offering students a series of easy tasks can lead to a false sense of self-efficacy and can limit access to challenging mathematics. Ironically, research indicates that students need to experience periodic challenge and even momentary failure to develop higher levels of self-efficacy and task persistence (Middleton and Spanias 1999). Achieving a balance between opportunities for success and opportunities to solve problems that require considerable individual or group effort requires teachers to design curricular materials and instructional practices carefully (Woodward 1999). In the following sections, we describe our efforts to develop persistent and flexible problem solvers by looking deeply at instructional practices and mathematical tasks.

Context of Our Classroom Design Research through Lesson Study

The design research process enabled the teachers and researcher to document the Lesson Study, to develop sequences of instructional strategies and tools, and to analyze student learning and the means by which that learning was supported. During the academic year, the classroom teachers and the researcher collaboratively planned four problem-driven lessons focused on developing persistent and flexible problem solvers. Before each lesson, the teachers and researcher met to discuss the lesson objective, important mathematics, the design of the mathematical task, and the expected flow of the lesson. Also, the group spent considerable time discussing students' anticipated responses and common misconceptions in an effort to develop conceptual supports that would scaffold the tasks for diverse learners. We collected data to document the design process through the pre- and postlesson discussions and through artifacts such as lesson plans, task sheets, and conceptual supports. When teaching the lesson, each teacher had an observer who recorded notes on students' engagement, responses, and questions that elicited rich mathematical discourse. During postlesson meetings, the teachers and researcher met to discuss the outcome of the lesson, evidence of student learning, and how the lesson design contributed to developing persistent and flexible problem solvers.

Through each cycle, the teachers and researcher designed different pedagogical tools to support our research aim. One such tool was a set of prompts that encouraged learners to reflect orally or in writing about their use of communication, flexible thinking, and persistence as they approached mathematical tasks (fig. 12.1). These prompts emerged from collaborative discussions among the Lesson Study teachers about the research goals. Before the school year began, we discussed what characterized persistent and flexible problem solvers who communicated clearly and respectfully. We designed the prompts to share our vision with our students and help them internalize these traits. The chart was organized so that teachers could have students respond to a particular column or row or give students choices about which prompts to use. This reflection not only gave the students insight into their problem-solving process but also gave the teacher–researchers a window into students' motivation, persistence, and flexibility in thinking.

Clear Communication	Respectful Communication	Flexible Thinking	Persistence
What math words could help us share our thinking about this problem? Choose 2 and explain what they mean in your own words.	Did someone else solve the problem in a way you had not thought of? Explain what you learned by listening to a classmate.	What other problems or math topics does this remind you of? Explain your connection.	What did you do if you got stuck or felt frustrated?
What could you use *besides words* to show how to solve the problem? Explain how this representation would help someone understand.	Did you ask for help or offer to help a classmate? Explain how working together helped solve the problem.	Briefly describe at least 2 ways to solve the problem. Which is easier for you?	What helped you try your best? *or* What do you need to change so that you can try your best next time?
If you needed to make your work easier for someone else to understand, what would you change?	What helped you share and listen respectfully when we discussed the problem? *or* What do you need to change so that you can share and listen respectfully next time?	What strategies did you use that you think will be helpful again for future problems?	Do you feel more or less confident about math after trying this problem? Explain why.

Fig. 12.1. Reflection prompts to encourage persistence and flexibility in problem solving

In the following sections, we share classroom accounts that describe students' development of persistent and flexible thinking. The participating teachers taught fourth- and fifth-grade students of diverse ability levels, including students with individualized education plans, students of average ability, and gifted-and-talented students.

Lessons That Elicited Students' Persistence and Flexibility in Problem Solving

Through our research lessons, we discovered essential design features that elicited students' persistence and motivation for solving problems. One design element involved presenting rigorous mathematical tasks that challenged students' thinking and required justification and reasoning. In a lesson called "Possible Solution Set," we posed a task where students found all the possible ways to have a three-digit house number whose digits had a sum of 12. In addition to discovery of number combinations, the underlying problem-solving focus was to encourage students to use a table or an organized list to keep track of number combinations. For an extension, we asked students to find all the three-digit house numbers whose digits had a product of 24. Once students discovered a mathematical strategy, giving them related problems or classes of problems was important so that they could transfer their strategy development to other problem types.

Through this lesson, students developed persistence as they worked and discovered multiple answers that satisfied the criteria (fig. 12.2). Several students began by listing random combinations of numbers, but through collective inquiry, those students soon realized that their method was not efficient for keeping track of all the number combinations. In fact, once they noticed their classmates' using a variety of strategies, such as a table, a tree diagram, or an organized list, these students developed an appreciation for different problem-solving strategies. As the teachers circulated through the room asking for solutions, students could see that several ways to solve the problem existed. Certain students were also better at verbalizing their strategy and thinking processes, whereas others created excellent tables and organized lists. We noticed that students were motivated to find all the possible combinations. When we asked, "How do you know that you have all the number combinations?" students had to prove their thinking. This emphasis on justification reinforced the importance of persisting until students were certain that they had solved the problem. One student commented, "I feel more confident after doing this problem because I really get stuck on knowing when to do an organized list, but now I know when to make one."

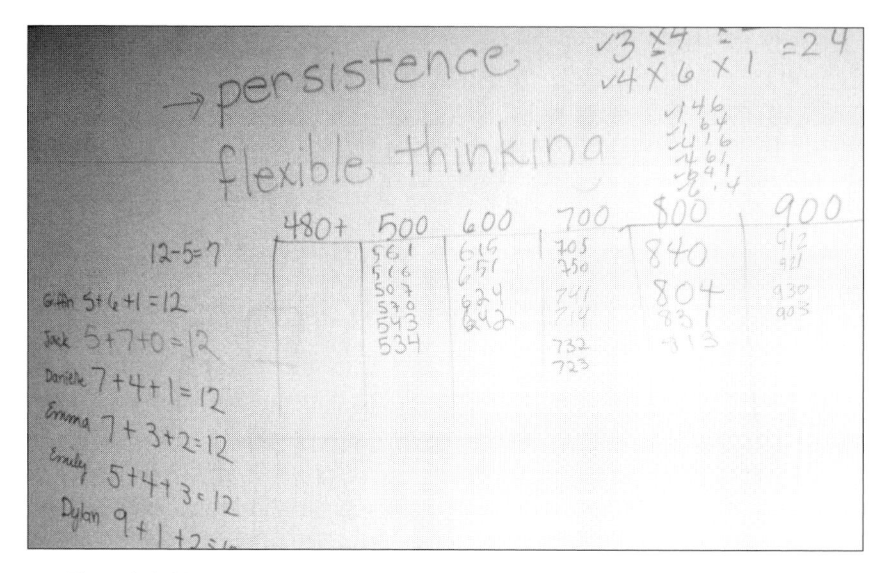

Fig. 12.2. Finding all the combinations by using a variety of strategies

We created another design element that used tiered tasks that allowed for multiple entry points, engagement, and differentiation. We used tiering to adapt an NCTM Illuminations lesson about the properties of triangles (see illuminations .nctm.org/LessonDetail.aspx?ID=U191). For tiered lesson one, "What Can You Build with Triangles?" we explored ways of building different basic shapes from triangles to investigate the properties of a triangle and the relationships among other basic geometric shapes. For tiered lesson two, "What's Important about Triangles?" students explored relationships among the side lengths of a triangle to determine whether they could construct triangles from these lengths. For an extension and a challenge, in tiered lesson three, "How Many Triangles Can You Construct?" students identified patterns in Sierpinski's triangle and built a foundation for understanding fractals. The tiered lessons enabled students to work with worthwhile tasks that were neither too easy nor too difficult. Our rationale was based on our understanding that tasks that are too easy may bore students, whereas tasks that are too difficult may frustrate them.

In a fourth-grade classroom, the research lesson focused on tiered lesson two, "What's Important about Triangles?" which explored relationships among the side lengths of a triangle. Students determined whether they could construct triangles from given lengths. This investigation allowed students to become mathematicians who constructed and tested conjectures by first predicting whether the measurements of the side lengths would make a triangle. Then they constructed the triangle to test their predictions (fig. 12.3). Students could look for commonalities among the measurements that worked compared to the

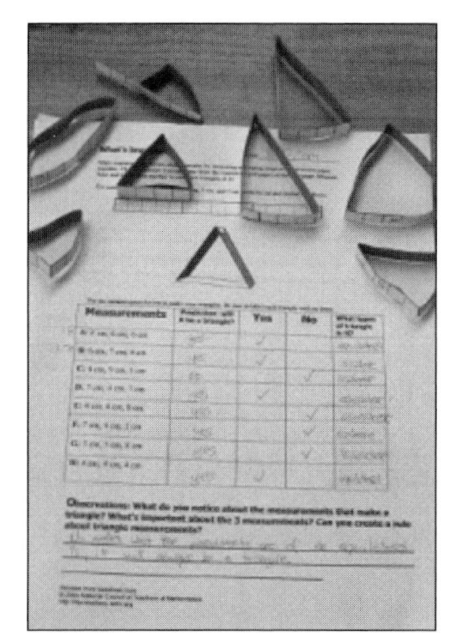

Figure 12.3. Making and testing conjectures about side lengths of a triangle

lengths that did not yield a triangle. This inquiry approach ignited students' curiosity to understand why they obtained their results. As students collectively immersed themselves in this mathematics inquiry, several started to articulate conjectures, and the teacher had students share their conjectures on the board to develop collective knowledge. For example, Shailyn and Sebastian stated that "If two of the [smaller] sides are added and are smaller than the third, you cannot make a triangle." Building on this conjecture, Janae stated, "If the two smaller sides are added and they are bigger than the third, then you can make a triangle." After giving a few examples of measurements and noting "yes" or "no," Jimmy added that "All equilaterals work." The generalizations resulting from individual and shared conjectures mirrored the progression of important mathematical ideas that mathematicians have made throughout history. The shared investigation and community of math inquiry was motivating for students as they discovered properties of triangles.

In a fifth-grade class of advanced students, one teacher focused on tiered lessons two and three. Through the two different levels, all students explored the patterns that occur when triangular pyramids iterate. Through their assigned group discovery activity, students working on lesson two, "What's Important about Triangles?" found exactly how various kinds of triangles are formed and a rule to generate all triangles. Those working on lesson three, "How Many Triangles Can You Construct?" (Kelley 1999), discovered a pattern in the formation of the Sierpinski triangle to determine a rule for its structural iteration. Students drew several iterations of the Sierpinski triangle and then used a computer program to see dynamically how the triangle continues to generate the repetitive pattern. The seemingly buried commonality between both exercises was the need to use emerging patterns to unlock the mysteries behind geometry's strongest and most-studied shape. The rule that the lengths of the two smaller sides of a triangle must have a sum greater than the length of the largest side prodded the others to conclude that the Sierpinski iteration worked in accordance with ascending exponents at every new generational level.

The true test of persistence was building the Sierpinski triangle (fig. 12.4). The specific model that the students built required the prior construction of 256 triangular pyramids from a template. This activity was representational for this class of sixteen students because each stage required precisely sixteen triangular pyramids. Building the model as a class community helped foster the idea

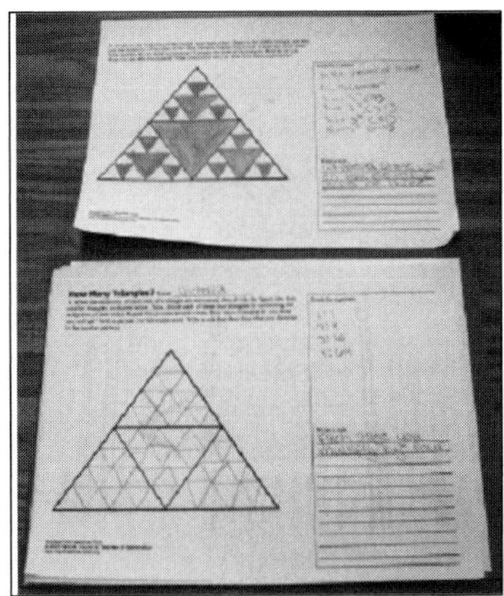

Figure 12.4. Students' development of motivation and persistence through building triangles

that the large-scale triangle was a shared project as well as a mathematical challenge. Mutual respect and collective effort motivated students to complete such a seemingly difficult exercise. The teacher set the tone and offered the nurturing assurance for the community of inquiry to mature. Closing with reflecting on the problem-solving prompts permitted students to share their thinking, but it also gave the teacher valuable clues for how to best facilitate follow-up and extension lessons.

Through students' responses from their reflection logs, we observed their development of a productive disposition toward mathematics. When asked what they did if they got stuck or felt frustrated, students responded, "We asked for help, and we tried to look at things in a different way" and "I asked for help and offered help. I think working in groups is easier because two people can do more than one." When asked what they could use other than words to show how to solve the problem, students responded, "I believe diagrams trigger people's minds so they understand and visualize the problem better" and "If you find the rule and the pattern, you can better see how a problem works." The research team collected and categorized students' comments according to our research aim of evidence of persistence and flexibility in thinking (table 12.1).

Table 12.1
Evidence of students' development of persistence and flexibility in problem solving

Persistent problem solver	**Flexible problem solver**
"I feel much more confident in math, because this problem showed me different problems, strategies, and persistence. The persistence helped me because I put my mind to it." —Alex	"Using the formula to predict if the sides would make a triangle helped me a lot. It is a good strategy for the future." —Sam
"What helped me try my best was when Michael didn't understand something and made me know I had to try harder to explain it better." —Liam	"This problem reminded me of the shapes that we made with the straws and twist ties." —Danielle
"I felt more confident about math after trying this problem because I proved to myself that if I am persistent, then I can accomplish things in math that I set my mind to." —Lauren	"I like trial and error because you start with a big guess and narrow it down." —Griffin
	"A strategy that will help me in the future would be the rule that we found out today." —Emma

Table 12.1—*Continued*

Persistent problem solver	Flexible problem solver
"I feel a lot more confident about math after those problems because I know what it feels like to be persistent, and I like it! So I'm going to keep going for that feeling." —Emily	"A strategy that I would use again after this problem would be guessing. I think this because many problems involve estimating. I'm guessing more and doing it better." —Alex
"What helped me to do my best was the hard questions. The more confusing it was, the more I liked it to try my best." —Liam	"I think that doing the number sentences will help me in the future." —Molly
"I just kept going like a snow plow stuck in the road. I didn't wait for the spring to come. I kept going." —Griffin	"This reminds me of when we tried to find perimeter in the beginning of the year. When we first did this, we could barely multiply and divide." —Liam

Core Instructional Practices for Developing Persistent and Flexible Problem Solvers

As we used research lessons to analyze our teaching, we identified four core instructional practices that were instrumental in establishing classroom norms that had a positive effect on students' disposition toward mathematics (fig. 12.5). These four core instructional practices complemented each other and converged to build a safe and stimulating classroom environment that nurtured a community of mathematics inquiry.

First and most important was establishing a community of mathematics inquiry that embraced challenges. Students were more motivated when they felt that they were part of a vibrant and rigorous learning community. Classroom norms that encouraged persistent and flexible problem solvers took time to build. These norms included attributing value to struggle, respecting diverse strategies, communicating mathematical ideas, seeking and giving help when students got stuck, evaluating different strategies for their advantages and disadvantages, self-correcting, being flexible enough to change one's ideas to garner further mathematical insight, and placing value on being a good problem solver.

Second, the emphasis on respectful and clear mathematics communication allowed students to engage in rich, in-depth mathematics argumentation with

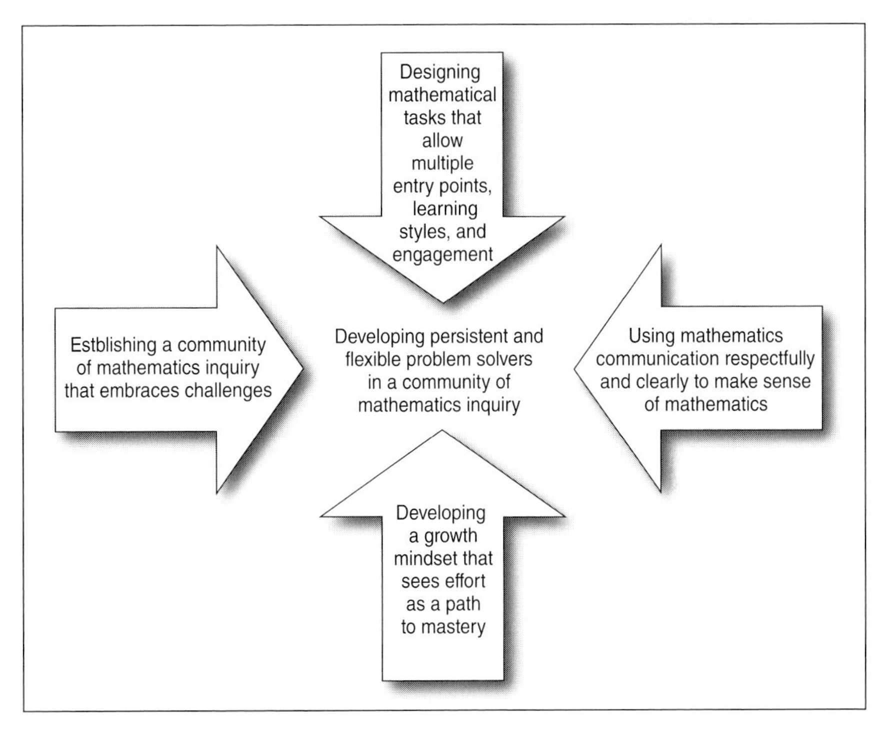

Figure 12.5. Four practices to establish classroom norms to develop persistent and flexible problem solvers

reasoning and proof, just as mathematicians do. Students learned to express their ideas by using multiple tools such as drawings, models, words, and numbers to convince one another, elaborate on each other's ideas, and translate among representations. Mathematical communication included in-depth discussions when students shared and compared strategies, verbalizing the metacognitive process that explained why one pursued a worthwhile strategy while abandoning inefficient ones, determining what questions to ask when one got stuck, and permitting students to defend their answer to build confidence in their reasoning. At times, giving students the space and time to respectfully argue mathematical ideas and convince one another gave reluctant learners and ones who needed more scaffolding the opportunity to make sense of the mathematics. Viewing wrong answers as partially correct and reflecting on finding the part that is wrong and understanding why it is wrong can be a powerful aid to understanding and promoting metacognitive competencies.

Third, designing meaningful mathematics tasks that accommodated multiple entries, learning styles, and engagement was the instructional backdrop for the described activities to happen in the mathematics classroom. As teachers, we learned to use questions to guide and coach our students and to know when to intervene and when to let students grapple with the problem. We modeled the use of Polya's (1957) steps to problem solving by verbalizing the self-monitoring process that is vital during problem solving. We discussed persistence and flexibility and how those ideas apply to problem solving in mathematics. We illustrated what persistent students and their work looked like and then reinforced these behaviors when students demonstrated those dispositions. We likewise modeled and acknowledged examples of flexible thinking. Research calls this modeling "cognitive modeling," in which one verbalizes one's metacognitive processes when solving problems (National Research Council 2004, p. 241). Modeling, acknowledging, and highlighting student behaviors that we wanted to see was important. Those behaviors were visible not only in action but also in students' work.

Finally, developing a growth mindset that views effort as a path to mastery was integral for not only the students but also the classroom teachers. To make this an explicit expectation, the researcher developed an assessment rubric to evaluate students' progress in developing mathematical proficiency through the school year on the basis of demonstrated effort in these areas (fig. 12.6). Using this rubric, teachers noted students' development of productive dispositions toward mathematics along with the other important strands of mathematics proficiency, such as conceptual understanding, procedural fluency, strategic competence, and adaptive reasoning. Giving frequent feedback helped students recognize their progress in learning and gave them chances to do even better, which was motivating.

The teachers also supplied an exit pass so that students could self-assess their effort and reflect on their learning (fig. 12.7).

Helping students learn to appreciate multiple approaches to problem solving gave students an appreciation for the flexibility in thinking required to solve complex problems. Modeling reflection on the problem and discussions of various means and methods to solve the problem (manipulatives, charts, diagrams on centimeter paper, looking for patterns, and finally searching for a rule) created a foundation for the students' "problem attack." We recorded these strategies and structures on the board, as well as the term *persistence* and what it looks like in student behaviors. Students could refer to these visual clues for plans of attack if they thought their plan might need to be rejected and another plan substituted. This repertoire of strategies gave students alternative pathways to find solutions (kinesthetic, visual, and auditory). In later discussions, most students felt that diagrams and charts helped them find patterns and that patterns led to full solutions. They supported the theory that building the problem with manipulatives, and then recording their work as diagrams, led to confidence in their solutions.

Assessing Mathematical Proficiency
Activity________________________ Date: ____________

Student Name	Effort
Productive Disposition	
Tackles difficult tasks	
Perseveres	
Shows confidence in own ability	
Collaborates/shares ideas respectfully	
Strategic Competence	
Uses strategies flexibly	
Formulates and carries out a plan	
Creates similar problems	
Uses appropriate strategies	
Communication of Reasoning and Proof	
Justifies responses logically	
Reflects on and explains procedures	
Explains concepts clearly using the language of mathematics	
Conceptual Understanding	
Understands the problems or tasks	
Makes connections to similar problems	
Uses models and multiple representations flexibly	
Procedural Understanding	
Uses algorithm properly	
Computes accurately	

Scoring Rubric
3: Secure (Student demonstrates effort consistently.)
2: Developing (Student demonstrates effort most of the time.)
1: Beginning (Student demonstrates effort some of the time.)
0: Not demonstrated (Effort not demonstrated.)

Fig. 12.6. Assessing mathematical proficiency

EXIT PASS

Today, I put effort in my math thinking and math work. ☺ 😐 ☹
I learned that . . .
I still need help on . . .

Fig. 12.7. Exit pass

Conclusion

Before teachers can create classroom norms that foster persistent and flexible problem solvers, the teacher and students must agree on some beliefs and behaviors about teaching and learning. All members of the mathematics community need to commit to developing a growth mindset (table 12.2). One of the biggest commitments that teachers must make when building a community of mathematics inquiry is to give students the time and space to grapple with meaningful mathematics investigations. Students will at first feel frustrated and not know where to begin if they are accustomed to teachers' spoon-feeding them through problems. However, teachers need to become comfortable with this feeling and realize that this is the first stage of problem solving. As students persevere through problems, they will gain an appreciation for solving problems and learn how to make sense of mathematics.

Table 12.2
Commitment to a growth mindset and building a community of mathematics inquiry

Students' commitment to a growth mindset	Teachers' commitment to growth mindsets
I will persevere through problems and be productive. "Stick with it!" attitude.	I will give students time and space to grapple with problems and validate their efforts and persistence. I will distribute practice over time and give challenging problems.
I will make sense of mathematics through my written work and my participation in discussion as we work together.	I will choose meaningful and productive tasks and guide students' effort in learning important mathematics and building collective knowledge.
I will consider multiple strategies to learn the most efficient ways to approach a problem.	I will anticipate students' responses and elicit, support, and extend students' thinking.

Through the Lesson Study professional development, teachers also grappled with mathematics problems and learned the value of persistence and flexibility in thinking when solving problems. This experience was essential for teachers to understand the importance of having a productive disposition toward mathematics. Taking this experience and translating it into classroom practice, teachers recognized that changes in beliefs and behaviors needed to start with them. Through Lesson Study, we continued to focus on our overarching goal, which helped us be intentional in developing classroom norms that would facilitate this change. Enabling children to develop persistence and flexibility boosted their self-confidence and helped them embrace the importance of mathematical thinking.

REFERENCES

Dweck, Carol. *Mindset: The New Psychology of Success*. New York: Ballantine Books, 2006.

Elliot, Andrew, and Carol Dweck, eds. *Handbook of Competence and Motivation.* New York: Guilford Press, 2005.

Kelley, Paul. "Build a Sierpinski Pyramid." *Mathematics Teacher* 92 (May 1999): 384–86.

Middleton, James, and Photini Spanias. "Motivation for Achievement in Mathematics: Findings, Generalizations, and Criticisms of the Research." *Journal for Research in Mathematics Education* 30 (January 1999): 65–88.

National Council of Teachers of Mathematics (NCTM). *Curriculum and Evaluation Standards for School Mathematics*. Reston, Va.: NCTM, 1989.

National Research Council. *How Students Learn Mathematics in the Classroom.* Washington, D.C.: National Academies Press, 2004.

Olkin, Ingram, and Alan Schoenfeld. "A Discussion of Bruce Reznick's Chapter." In *Mathematical Thinking and Problem Solving,* edited by Alan Schoenfeld, pp. 39–51. Hillsdale, N.J.: Lawrence Erlbaum Associates, 1994.

Polya, George. *How to Solve It*. Princeton, N.J.: Princeton University Press, 1957.

Schoenfeld, Alan. *Assessing Mathematical Proficiency.* New York: Cambridge University Press, 2007.

Weiner, Bernard. "Motivation from an Attributional Perspective and the Social Psychology." In *Handbook of Competence and Motivation*, edited by Andrew Elliot and Carol Dweck, pp. 73–84. New York: Guilford Press, 2005.

Woodward, John. *Self-Concept, Self-Esteem, and Attributional Issues in Secondary Math Classes.* Presented at the Pacific Coast Research Conference, San Diego, 1999.

Chapter 13

Mark's Development of Productive Disposition and Motivation

Hope Gerson
Charity Hyer
Janet Walter

University calculus students were working to understand average and instantaneous velocity by analyzing photographs of a cat going from a walk to a gallop (fig. 13.1). The instructors had not introduced the idea of derivative because the students were building the idea themselves. After working on the task for three days, one student, Mark, came in before class, bubbling with excitement, intent on sharing with the instructors the insights that he had found by reading his textbook (Garner 2006). He then redeveloped and justified the definition of derivative, on the board, without using his textbook.

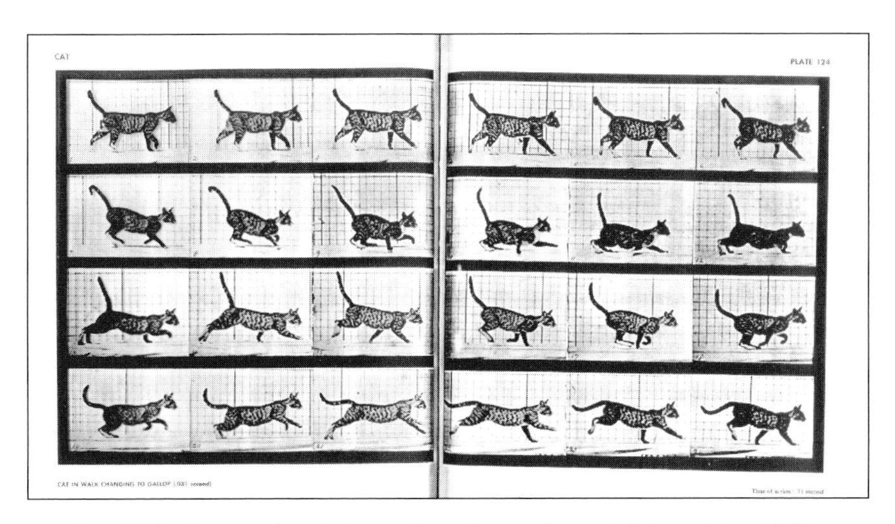

Fig. 13.1. Sequential photographs of a running cat (Muybridge 1887)

We found this episode to be compelling for several reasons. First, Mark chose to read the textbook on his own, without prompting from the instructors. Second, Mark redeveloped the definition of derivative by using reasoning and sense making rather than memory. He used the textbook to develop conceptual understanding rather than reading the examples to solve stock problems. Third, Mark's enthusiasm about what he found in the textbook was unusual—he called it "invigorating." In our experience, the typical response to assigning students to read about derivatives in the textbook is a collective groan.

Thompson (1994) suggested that case studies of individual students are important for understanding how students think, act, and develop understanding. Many studies point to failures, struggles, or misconceptions that students may have, but fewer case studies highlight student successes (Bezuidenhout 2001; Davis and Vinner 1986; Szydlik 2000). What would it be like if students were motivated to search their textbooks for meaning and understanding—beyond procedures highlighted by a blue box—and how could students develop self-efficacy, productive disposition, and motivation while engaging in rich tasks?

Background and Context

We first define and give background on self-efficacy and productive disposition. Next, we share two contexts, agency and intellectual need, that will be helpful in our discussion of self-efficacy and productive disposition. We then explain the setting where this compelling episode occurred and the episode itself, in which Mark and his group's previous mathematics experience played a role. We also examine what this episode can teach us about how this setting might help students develop self-efficacy, productive disposition, and motivation to learn difficult mathematics with conceptual understanding.

Self-Efficacy and Productive Disposition

Self-efficacy (Bandura 1989) and productive disposition (Kilpatrick, Swafford, and Findell 2001) are terms to characterize students' attitudes and beliefs about mathematics and their capacity to learn and do mathematics. Students' attitudes about mathematics and belief in their capacity to do mathematics strongly affect their motivation. Bandura defines self-efficacy as one's perception of one's own capability. People's self-efficacy determines how well they deal with mistakes, false starts, and failures and how much they will persist in solving a difficult mathematics problem. People who have a strong sense of efficacy will believe that they can control their own situations.

Bandura (1989) further suggests that developing resilient self-efficacy requires experience in mastering difficulties through persistent effort. If people experience only easy successes, they come to expect quick results and failure undermines their sense of efficacy. Some setbacks and difficulties in human pursuits are useful by teaching that success usually requires sustained effort. After people become convinced that they have the skills and knowledge to succeed, they persevere in the face of adversity and quickly rebound from setbacks. By sticking it out through tough times, they emerge from adversity with a strong sense of efficacy (Bandura 1989, p. 5).

The National Research Council (in Kilpatrick, Swafford, and Findell [2001]) defines productive disposition as "the tendency to see sense in mathematics, to perceive it as both useful and worthwhile, to believe that steady effort in learning mathematics pays off, and to see oneself as an effective learner and doer of mathematics" (p. 131). The *Curriculum and Evaluation Standards for School Mathematics* (National Council of Teachers of Mathematics [NCTM] 1989) further defines seven characteristics of productive disposition in mathematics: seeing mathematics as both useful and worthwhile; believing that mathematics makes sense; persisting in solving difficult problems; perceiving oneself to be capable of doing mathematics (self-efficacy); solving problems flexibly; being interested, curious, and inventive in mathematics; and reflecting on mathematical thinking.

Role of Agency and Intellectual Need

Speiser, Walter, and Maher (2003) make a convincing case that students need "time, freedom, and diverse personal experience" to build mathematics understanding. We define agency in mathematics learning as "the requirement, responsibility, and freedom to choose based on prior experiences and imagination, with concern not only for one's own understandings of mathematics, but with mindful awareness of the impact one's actions and choices may have on others" (Walter and Gerson 2007, p. 209).

Students' choices largely form their beliefs about themselves. Therefore, agency plays an important role in developing self-efficacy. Bandura (1989) suggests that students' beliefs in their ability to control a situation are related to their actual ability to control the situation. When students succeed, their self-confidence increases. When that success stems from choices that they made, the students' confidence in their choices also increases.

Harel (2007) defines intellectual need as "a behavior that manifests itself internally with learners when they encounter an intrinsic problem—a problem they understand and appreciate" (p. 274). An interesting and engaging problem that a student wants to solve activates that student's intellectual need. Harel suggests that intellectual need is indispensable for students to create new mathematical understanding. Harel further suggests that intellectual need is firmly tied to epistemological justification, or "the learner's discernment of how and why a particular piece of knowledge came to be" (Harel 2008, p. 488).

Agency is omnipresent; no matter how constrained, a student can always choose a course of action and ultimately is responsible for his or her own learning. Therefore, the teacher cannot develop agency, but a teacher's actions can constrain or encourage it (Walter and Gerson 2007, p. 209). Whereas agency resides within an individual, the instructor can, by selecting appropriate and meaningful tasks or problems, establish student intellectual need. When students choose to engage in problem solving and create their own strategies for solving problems, the mathematics that they develop becomes personally meaningful to them (Castle and Aichele 1994) and they become more effective problem solvers (Siegler 1996). Both agency and intellectual need are tied to motivation and to developing self-efficacy and productive disposition.

The Setting

In winter 2006, Janet Walter and Hope Gerson team-taught an experimental Calculus I course at a large, private university in the western United States. The class met two hours per day, three days per week, for fourteen weeks. The twenty-two students sat in groups of four to five per table and collaborated on rich open-response mathematical tasks that the instructors gave them.

The instructors' belief that exercising personal agency is necessary for learning (Walter and Gerson 2007; Brown 2005) strongly influenced the course design. To maximize students' exercise of agency, we thought that students should drive the mathematics that the course covered. Therefore, the instructors did not give students prior instruction on important calculus content. Instead, the instructors expected them to solve open-response mathematical tasks from which they could build important calculus content, share their ideas with one another in a supportive atmosphere that valued understanding others' thinking more than criticizing others' thinking, and use their own and others' mathematical ideas to further their own learning of mathematics. In this design, students exercised agency by choosing what mathematical content to explore in the tasks, the prior knowledge they could use as building blocks to create new knowledge, what questions and hypotheses to make, how to create and test mathematical ideas, what solution strategies to use, how to evaluate their own and others' solution strategies and mathematical ideas, and what was important to include in a write-up of each task.

The instructors wrote some tasks, whereas others came from several sources, including research, reform calculus texts, and traditional calculus texts. The chosen tasks were (*a*) rich in mathematics content, in that each task could elicit many different important calculus ideas; (*b*) open-response format, in that they did not offer clues to what solution strategy to use; and (*c*) epic, in that students spent from two to twelve class days working on each task. Also, most of the tasks were in context so that students could use prior knowledge of their world and common sense to help make sense of and solve the problems. To solve the tasks, students would have to grapple with and develop conceptual understanding of important calculus content.

As mentioned, the instructors did not offer prior instruction on calculus content. Student choices drove the content of the course. Instructors would come to the board occasionally to present conventional notation or vocabulary, to facilitate the transitions between tasks or presentations, or to handle administration of the course. For most of the class time, the instructors listened to students to understand how they were thinking about a task, what mathematical ideas they were using, and what important calculus ideas were emerging. Sometimes the instructors stood by a table listening and watching, and other times they would ask questions to clarify student thinking. During presentations, the students and instructors questioned the presenters to fully understand the models, methods, and solutions. The instructors also facilitated the development of social norms by modeling listening and questioning techniques among the groups and during presentations.

Participants

The data in this article consist of 6.5 hours of videotaped collaborative work in small groups, students' written work, an initial written survey, and a clinical interview from the end of the semester. We focus here on Mark, a student in the class, and his collaborative work with his group. An engineering major, Mark had taken calculus in high school. When we asked him about his previous calculus experience, he said that he was "a very mediocre student that was more concerned with sleeping, swimming, and girls than school" and slept through most of his calculus class. He learned procedures, "enough to get by," but never really understood "what was going on." After taking a two-year break and returning to the university, he felt unprepared for college-level mathematics. He enrolled in an intermediate algebra course and then took college algebra and trigonometry. Mark interacted in class with the other members of his group: Chris, Kam, and Josh. Chris, a political science major, and Josh, an economics major, had both taken calculus previously. Kam was an engineering major taking calculus for the first time.

The Cat Task

Our analysis will focus on the second task, the Cat Task (Speiser and Walter 1994), which appeared at the end of the second week of class. Before the Cat Task, the students worked on the Desert Motion Task (diSessa et al. 1991), in which students read the following: *A motorist is speeding across the desert on a long, straight stretch of highway. She is very thirsty and happens to catch a glimpse of a cactus with a water faucet sticking out from its side. She is traveling at such a high rate of speed that she overshoots the cactus, so she stops the car, backs up, jumps out of the car, gets a drink of water from the cactus, and then resumes her drive. How might you represent the motion of the car?* Working on the Desert Motion Task, students developed ideas about displacement, velocity, and acceleration, but they were still negotiating conceptual understanding of these ideas, their representations, and the relationships among them. The word *derivative* had been used—but not defined—and not always correctly. Students were still forming ideas about motion, vocabulary, and notation when they began the Cat Task.

Quantifying the Data

Mark and his classmates received a copy of the stop-motion cat photos (fig. 13.1) along with the following instructions: *Based on the information you can gather from the pictures of the cat, how fast is the cat moving when photographed in Frame 10? How fast is the cat moving when it is photographed in Frame 20? [The background grid lines are 5 cm apart, every tenth line is a darker line, the time interval between frames is 0.031 seconds, the elapsed time for all 24 frames is 0.71 seconds.]*

Mark's first approach was to draw a graph of the cat's displacement over time (fig. 13.2). He worked meticulously for about fifty minutes taking careful measurements, only occasionally paying attention to the discussion of his group. The graph helped Mark organize the data so that he could better understand and theorize about the cat's motion. He determined that the cat was increasing speed in each frame, and he could discuss the relative speed of the cat in each frame. After almost an hour (fifty minutes, thirty seconds [50:30]) of working mostly independently, Mark took an interest in the discussion around him and engaged in a conversation stemming from the data points that Josh plotted on his calculator. In this discussion, Josh introduced the word *derivative*. The time that elapsed before anyone in Mark's group introduced the term indicated that the students did not recognize immediately that the derivative is related to instantaneous velocity—even though they had previously built this connection in work on the Desert Motion Task.

50:32	*Mark:*	Position over time. Now if I remember [from the Desert Motion task], the slope of the position line is velocity.
	Josh:	So if you take the derivative, which means the slope of that line [*pointing to Mark's graph*] . . .
	Mark:	There you go, derivative at each point.

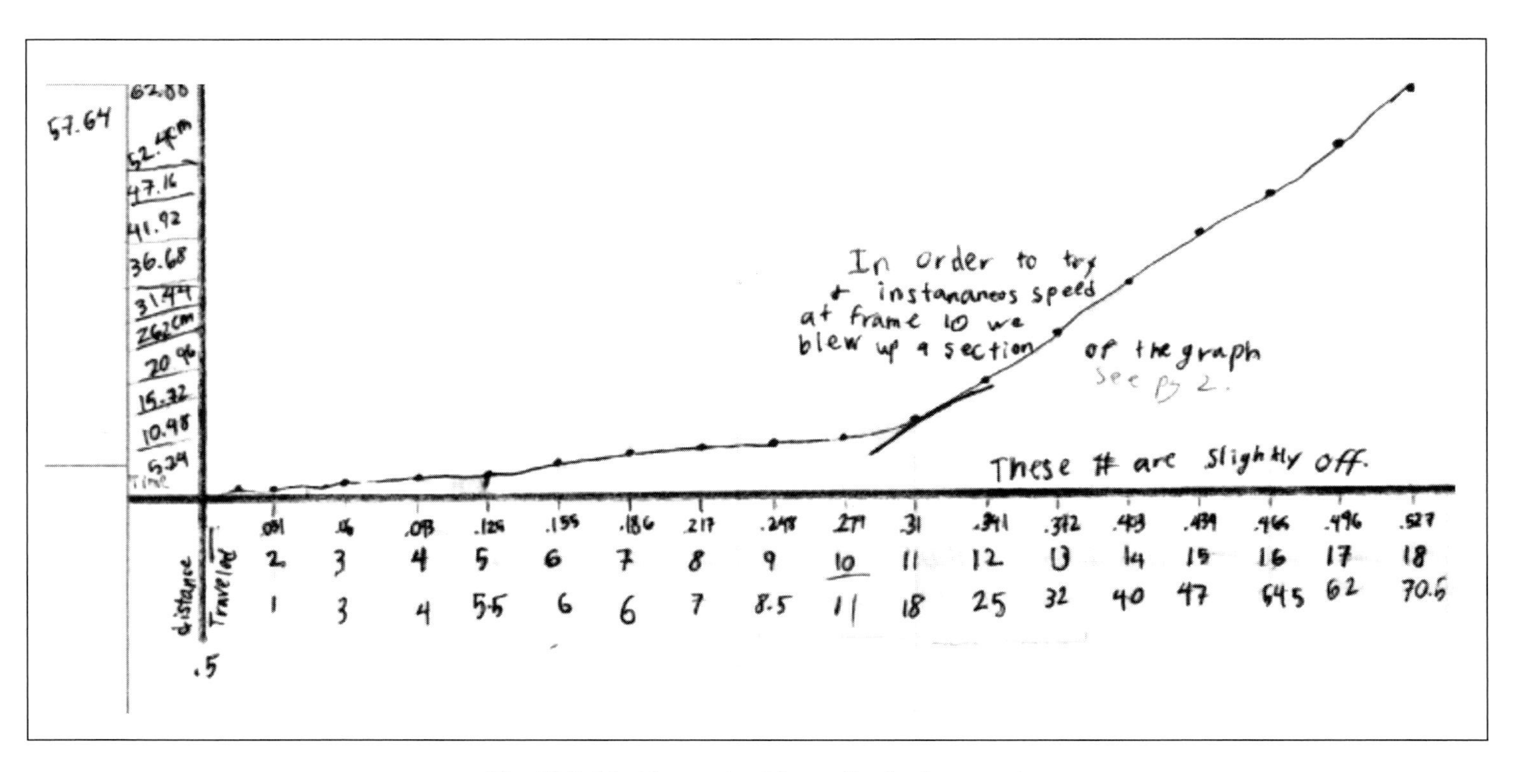

Fig. 13.2. Mark's graph of the cat's displacement

Josh: If you take the derivative of that line, you're going to get the velocity graph.

Trying to Find Instantaneous Velocity

To develop the concept of derivative, the students moved to a discussion of slope. The students worked to conceptualize the slope for the first forty minutes of the second day. Mark knew that they needed two points to determine slope. He also knew that to find the slope of a line at a point, or at least an approximation, the two points must be very close together, or as he said, "infinitesimally close together." Mark turned his attention to finding these points and the slope between them.

1:43:57 *Kam:* Should we just, like, get [the graph] bigger?

Josh: Just keep zooming in?

Mark: That's the idea. You get, like, two points that are infinitesimally close together, [and] then you find the slope between those two infinitely close-together points. That's the whole idea between limits and stuff. But I never understood a derivative to begin with, like in high school, so . . . But we would still—if we want to blow up this section of the graph right here [*data points around frame 10*]—you know, like, we'd still need to know, with our archaic way of graphing things, we'd still need to know multiple points in between frames.

Mark and his group decided that the cat would probably have fairly fluid motion, and so a curve passing through the points should generate a good estimation of the cat's motion between the discrete data points. Mark saw the limitations of the discrete data that they had and how such data were limiting their estimates of the cat's velocity. Mark began to search for an equation that could model the discrete data.

Josh finally found an equation by using the regression function on his calculator and showed the graph on his calculator to the group about an hour and a half into the second day of the cat task (2:27:41). The group was interested in the meaning of Josh's graph and turned attention to making sense of the graph. Mark asked, "What's x in this equation? x is the time, and y is equal to the distance traveled, right?" Once Mark could determine the meaning of the variables x and y, he could use Josh's equation to find the velocity of the cat. A few minutes later Mark said, "So we might take, like, 39 and the next point up is, like, 39.00000001, but you know, you still, it's still an estimation" (2:35:30). So even plugging in two numbers very close together into Josh's equation is not exact enough for Mark. He said, "Like, it's gonna be, it's gotta be done with numbers or mathematically in the end. Cause you can't have infinite points, you know," (2:31:40). Here Mark indicated that he saw mathematics as useful and worthwhile, demonstrating one characteristic of productive disposition (NCTM 1989).

Thinking about the slope of the tangent "mathematically," Mark thought, would yield a more satisfying answer. By this time, Mark had thought deeply about tangent lines, slope, average velocities, continuity, and limits, thereby demonstrating interest, curiosity, and inventiveness—another characteristic of productive disposition (NCTM 1989). He was building the mathematics needed to understand the limit definition of derivative. Mark demonstrated his flexibility in problem solving, another characteristic of productive disposition (NCTM 1989), by using multiple representations (graphical, symbolic, and numerical) as well as considering multiple mathematical ideas in his solution of the Cat Task.

At this point Chris used the power rule that he had learned in a previous calculus class to take the derivative of Josh's regression equation. Mark and Josh were interested, but neither was satisfied with the procedural explanation that Chris gave of the power rule:

2:36:27	*Josh:*	Wait. With your calculator you took the derivative?
	Chris:	No, I just knew it.
	Josh:	How do you do it?
	Chris:	Um, the power rule. You take this exponent and you drop it down in front. Have you had calculus before?
	Josh:	It's been a long time, man!
	Chris:	OK, so the exponent goes down and you multiply it by the cofactor and then you take x to the zero is 1. One times that is the same, and then there's no x, so the constants go away.
	Mark:	Wait, could you show me what you just did?
	Josh:	But why? Why do you do that?

Mark and Josh wanted to make sense of the procedure. Chris's apparent expertise did not convince them. They wanted to understand the rule, thus demonstrating the characteristic of productive disposition (NCTM 1989) that mathematics make sense. By the end of class Mark was still dissatisfied with the answer. He rejected Chris's algorithmic solution and wanted to find something more accurate than they had before and that made sense.

Class presentations and the collaborative work of Mark's group interrupted his individual work. However, at the end of day three, Mark returned to his original solution and expressed his desire to take it further, thus demonstrating motivation and persistence, another characteristic of productive disposition (NCTM 1989).

5:00:13 *Mark*: I want to know if we can get something more accurate. That's my question—if we can get anything more accurate than what we have. Can it be done? If it can, I'm going to keep working at it.

That evening, Mark went home and looked up the definition of derivative in the book (Garner 2006). He was still looking for a more accurate solution that made sense, and he had Josh's regression equation to work with. He had established that he wanted to find the slope of the tangent line to the curve at a point. He knew that he needed two points to determine a slope but that the two points needed to be "infinitesimally" close. Thus, he had established not only an intellectual need but also the mathematical underpinnings necessary to understand the definition of derivative as the book presented it.

Reconstructing the Definition of Derivative

A half-hour before class on the fourth day, Mark came to show his instructors what he had learned from the book. When the instructors asked him to explain, he could do so, but he struggled through parts of his explanations—demonstrating that he was not just remembering; he was also reconstructing.

Mark was excited about the example that the book used. "They [the textbook] did it really cool. Like the problem they did. They did x^2." Mark explained how the book used h to indicate the distance between two x values. He made several mistakes determining pluses or minuses and what went where. When he carefully considered each part of the expression in the context of rise and run, he could correct the difference quotient and wrote

$$\frac{f(x+h)-f(x)}{x+h-x}$$

Next, Mark used the definition of derivative to find the derivative of $f(x) = x^2$. Throughout his twenty-five-minute presentation to the instructors, Mark showed how excited he was. He also showed that he was reflecting on his own understanding as well as the justifications that the book supplied, another characteristic of productive disposition (NCTM 1989).

5:33:32 *Mark:* And, they [the textbook] didn't give a very good proof of this, but it makes sense to me in my head. As h approaches zero, so then they just said that it equals $2x$ and that was the derivative of x^2 [*proudly showing off his work*].

Instructor: Cool.

Mark: Yeah, I was pretty excited when I—I was like, "Oh, it worked!"

Instructor: Why, why could they say that $2x + h$ is just $2x$, do you think? You said they didn't do a very good proof of it.

Mark: Well, in my mind, um, because it works in my mind because of this over here [*indicating his original diagram with secants and the tangent line*]. That's how it works in my mind . . . as h gets smaller and smaller and smaller and smaller, they say that zero is the limit for h. And it's almost as if we've actually reached the limit.

After his presentation to the instructors, Mark had a chance to teach the limit definition of derivative to his group members. When he explained it to them, he made none of the mistakes that he had made while explaining it to the instructors earlier and could justify each step and the various notations. When he drew his diagrams for his group, he was careful to label important points along the x- and y-axes, and his explanation of slope flowed easily from his diagram. Mark's explanation to his group members was much more concise than that to his professors. When group members asked questions, Mark was not flustered and could answer clearly, with strong justifications.

Mark retained his understanding of the derivative. In an interview at the end of the semester, Mark demonstrated a conceptual understanding of the definition of derivative by explaining the limit definition of derivative clearly. He could also correctly solve and justify his solution to a new problem involving average and instantaneous velocity.

Mark's explorations of the Cat Task helped him to create an intellectual need for mathematics that motivated Mark and his peers and made finding a solution "exciting." Mark said, "The difference is, like, I think it's 'cause we got frustrated because we didn't have it. We're, like, 'Aaarg! We want to use it! We want to figure out how to do it!'" Mark got frustrated while working on the task—but not frustrated enough to quit.

Discussion

This section describes Mark's evidence of a productive disposition and motivation, how the classroom structure affected his learning, and this situation's implications.

Productive Disposition and Motivation

Mark showed a productive disposition in several ways during his work on the Cat Task:

1. He saw mathematics as both useful and worthwhile.
2. He believed that math is understandable, not arbitrary.
3. When Chris showed his group the power rule for finding the derivative, Mark wanted to know why it worked and where it came from. Mark wasn't happy with a simple answer but wanted to make sense of mathematics and believed that it could and did have meaning.
4. He saw himself as an effective learner and doer of mathematics and that, with diligent effort, he was capable of learning.
5. He was interested and excited about learning the definition of derivative and exercised his agency to come to understand the concept of derivative on his own, outside class.
6. He reflected on his own and others' mathematical thinking, showing excitement over his own conceptual development of the definition of derivative.

Influence of Classroom Structure

Several key elements in the classroom structure supported Mark's development of productive disposition and motivation. Mark took advantage of ample time and opportunity to develop intellectual need, build conceptual mathematical underpinnings of the definition of derivative, and reflect on his understandings—all before reading the textbook. He exercised his personal agency to make choices that led to productive solution strategies, which helped him to pursue his interests in solving the problem and to develop self-efficacy and persistence. Allowing students to struggle for a week in class helped to make the solution much more meaningful to them as they generated intellectual need. This outcome expands Harel's (2008) suggestion that intellectual need stems from epistemological justification that the teacher supplies. Here, we see that intellectual need can also come from allowing students with diverse personal experiences to exercise personal agency over time as they engage together to solve a difficult problem.

It was not simply *reading* the textbook that enlightened Mark. He had tried to read the textbook more than once before his work on the Cat Task and had not understood it. A vital difference was that Mark's mathematical preparation developed over time while he tried to find the instantaneous velocity of the cat. To proceed with the problem, Mark organically developed the importance of slope, tangent lines, secant lines, continuity, and limits. His ideas became well defined through his struggles and communication. Mark's persistence in the face of frustration shows that he developed the self-efficacy that he lacked in high school.

When Chris expressed frustration, Mark described what he thought the instructors' intentions were: "I think they just, basically, we just have as much time until they're satisfied that we've come to an appropriate conclusion or until they've, like, or until we've exhausted all our options, and they're, like, 'Okay, what are you thinking? Guide your thought process.'" The fact that Mark believed that the teachers expected him to exhaust his possibilities may have contributed to his persistence. Also, Mark recognized the openness of the task when he used the words "an appropriate conclusion." Mark did not say "correct," "right," or even "*the* appropriate conclusion." This way of thinking is unusual. So many students would have believed that the teacher would just give them the answer when they had worked on the problem for a while. In this class the instructors did not constrain the students' agency in mathematical exploration. The solution to the problem was not the focus of this class. Rather, the focus was on developing conceptually important calculus ideas.

The classroom setting also allowed students to teach each other. They listened to each other and learned from each other, building their self-efficacy because they saw that they could be successful mathematics learners and teachers. Through their social interactions they could define terminology and notation. In their presentations to the class and their group collaborations, students clarified their thoughts and deepened their own understandings. Mark referred to the social interplay that helped him make sense of the task. He did not solve the task alone, yet his individual ideas greatly contributed to the collaborative efforts of his group. Mark's entire group was excited to solve the task and learn mathematics.

Implications

This paper examines the effort of one student who demonstrated both procedural and conceptual understanding of the definition of the derivative, retained his knowledge, applied it to new tasks, and enjoyed the whole process. He demonstrated productive disposition and motivation as he struggled to learn difficult and important mathematics. How might one replicate in other classrooms Mark's positive strides as he came to understand the definition of derivative? We believe that tasks, agency, time, and working with others were important characteristics to Mark's development of productive disposition and motivation to learn.

A teacher can give tasks that require the students to think deeply about important and difficult mathematics before or in place of direct instruction. A teacher can allot time for students to build and solidify ideas and explain why something is true or where it came from. The tasks in this article were rich in mathematics so that students could build several important ideas at once. For example, the Cat Task elicited discussions of average and instantaneous velocity, slope, derivative, limit, and continuity. So even though the class explored the task over many class periods, the students were learning a great deal of calculus in the process. Homework problems at the end of a section were another source of tasks. These problems, which students received before instruction, played a different role in learning mathematics. Rather than just affording practice, the tasks helped students to develop an intellectual need and curiosity about the mathematics that they were learning.

The teacher can set norms in the classroom that allow, and in fact require, students to exercise their personal agency regularly. We believe that this approach is vital to developing self-efficacy. As students make choices and discuss these choices with one another, they develop, justify, and solidify ideas. Creating such a setting can foster self-efficacy. Taken together, rich tasks, agency, time, and discourse employed in a nontrivial way foster productive disposition and motivation in the classroom, making learning fun and invigorating.

(This article is based on one author's master's thesis [Hyer 2007].)

REFERENCES

Bandura, Albert. "Human Agency in Social Cognitive Theory." *American Psychologist* 44 (September 1989): 1175–84.

Bezuidenhout, Jan. "Limits and Continuity: Some Conceptions of First-Year Students." *International Journal of Mathematical Education in Science and Technology* 31 (July 2001): 487–500.

Brown, Tony. "Shifting Psychological Perspectives on the Learning and Teaching of Mathematics." *For the Learning of Mathematics* 25, no. 1 (2005): 39–45.

Castle, Kathryn, and Douglas B. Aichele. "Professional Development and Teacher Autonomy," in *Professional Development for Teachers of Mathematics,* 1994 Yearbook of the National Council of Teachers of Mathematics (NCTM), edited by Douglas B. Aichele and Arthur F. Coxford, pp. 1–8. Reston, Va.: NCTM, 1994.

Davis, Robert B., and Shlomo Vinner. "The Notion of Limit: Some Seemingly Unavoidable Misconception Stages." *Journal of Mathematical Behavior* 5 (December 1986): 281–303.

diSessa, Andrea A., David Hammer, Bruce Sherin, and Tina Kolpakowski. "Inventing Graphing: Meta-Representational Expertise in Children." *Journal of Mathematical Behavior* (August 1991): 117–60.

Garner, Lynn. *Calculus*. 5th ed. Boston: Pearson Education, 2006.

Harel, Guershon. "The DNR System as a Conceptual Framework for Curriculum Development and Instruction." In *Foundations for the Future in Mathematics Education*, edited by Richard A. Lesh, Eric Hamilton, and James J. Kaput, pp. 263–80. Mahwah, N.J.: Lawrence Erlbaum Associates, 2007.

———. "DNR Perspective on Mathematics Curriculum and Instruction, Part I: Focus on Proving." *Zentralblatt fur Didaktik der Mathematik* 40 (July 2008): 487–500.

Hyer, Charity. "Discovering the Derivative Can Be 'Invigorating': Mark's Journey to Understanding Instantaneous Velocity." Master's thesis, Brigham Young University, 2007.

Kilpatrick, Jeremy, Jane Swafford, and Bradford Findell, eds. *Adding It Up: Helping Children Learn Mathematics*. Washington, D.C.: National Academies Press, 2001.

Muybridge, Eadweard. *Animal Locomotion*. Philadelphia: University of Pennsylvania, 1887.

National Council of Teachers of Mathematics (NCTM). *Curriculum and Evaluation Standards for School Mathematics*. Reston, Va.: NCTM, 1989.

Siegler, Robert S. *Emerging Minds: The Process of Change in Children's Thinking*. New York: Oxford University Press, 1996.

Speiser, Robert, and Charles Walter. "Catwalk: First Semester Calculus." *Journal of Mathematical Behavior* 13 (June 1994): 135–52.

Speiser, Robert, Charles Walter, and Carolyn A. Maher. "Representing Motion: An Experiment in Learning." *Journal of Mathematical Behavior* 22, no. 1 (2003): 1–35.

Szydlik, Jennifer Earles. 2000. "Mathematical Beliefs and Conceptual Understanding of the Limit of a Function." *Journal for Research in Mathematics Education* 31 (May 2000): 258–76.

Thompson, Patrick W. "The Development of the Concept of Speed and its Relationship to Concepts of Rate." In *The Development of Multiplicative Reasoning in the Learning of Mathematics*, edited by Guershon Harel and Jere Confrey, pp. 181–234. Albany, N.Y.: State University of New York Press, 1994.

Walter, Janet G., and Hope Gerson. "Teachers' Personal Agency: Making Sense of Slope through Additive Structures." *Educational Studies in Mathematics* 65 (June 2007): 203–33.

Chapter 14

Listening to Mathematics Students' Voices to Assess and Build on Their Motivation: Learning in Groups

Amanda Jansen

Efforts to enhance students' motivation should target the concerns and needs of students and the activities that students engage in. I believe that teachers should build on students' voices to guide instruction and increase its efficacy, because education is *for* students (Levin 2000). Inherent in this perspective is a belief that if teachers understood motivation from students' points of view, teachers could design increasingly productive experiences for students to learn mathematics.

Assessing Students' Motivation to Participate in a Classroom Activity

In this article, I share an investigation of twenty-four sixth-grade students' motivation to participate in small-group work in two mathematics classrooms. I interacted with the students when I worked with their teachers during a larger professional development project in which teachers were studying students' mathematical thinking to improve instruction. Two teachers asked me to help them learn about their students' motivation.

These two classrooms were at the same school in the mid-Atlantic United States. The student demographics were 50.6 percent African American, 27.5 percent white, 21.5 percent Latino/a, and 0.5 percent Asian American. According to district data, 69 percent of the students were from low-income families. Teachers used the *Mathematics in Context* (Encyclopaedia Britannica 2006) textbook materials, which were developed with funding from the National Science Foundation. The materials include rich mathematical problem-solving tasks with the potential to foster dialogue among students. Both teachers asked their students to work together on investigation problems from their textbooks in small groups at least once during each class period.

Selecting a Focal Classroom Activity

I first chose a focal classroom activity to support my learning about these students' motivation. I wanted to focus on one activity among students' experiences in mathematics classrooms, because one way to conceptualize motivation is as a motive (e.g., a wish, intention, or drive) to engage in a *specific* activity (Hulleman et al. 2008), in contrast to thinking about students' motivation to learn more generally. Students who are motivated to engage in a specific classroom activity could be more engaged during that activity and learn more from that activity as a result.

One salient step considered important for a focal activity was to identify an activity that was likely to support mathematics learning. I decided that participation in classroom discourse would be a worthwhile classroom activity, because engaging in discourse can help students develop their understanding of subject matter (Dillon 1994). At this point, I narrowed the activity further, because mathematics classroom discourse could occur in different structures, including whole-class discussions and small-group work.

When selecting a focal classroom activity, I was also concerned about engaging adolescents, since the students were in sixth grade. The National Council of Teachers of Mathematics (NCTM) writes about specific challenges associated with inviting adolescents to participate in classroom discourse in the Communication Standard for the grade 6–8 band in its *Principles and Standards for School Mathematics:* "During adolescence, students are often reluctant to do anything that causes them to stand out from the group, and many middle-grades students are self-conscious and hesitant to expose their thinking to others. Peer pressure is powerful, and a desire to fit in is paramount" (NCTM 2000, p. 268). I wondered whether adolescents would find small groups to be less threatening than whole-class discussions, because they would be sharing their thinking with a smaller group of peers.

Thus, I investigated students' motivation to participate in small groups. Talking with these sixth-grade students would allow me to learn whether the students encountered social concerns while working on mathematics with peers, as the *Principles and Standards* described. I believed that small-group work supported mathematics learning when students participated and tried to understand one another's thinking.

Considering What Productive Engagement Looks Like for That Activity

After selecting the focal classroom activity for assessing students' motivation, I reflected on what students might do if they engaged *productively* in small-group work. I conjectured that students' motivation to participate was related to how students participated or what they thought that participating meant. Students who did not express a strong motivation to participate in small-group work may not have been participating in a way that would help them benefit from it or see the value in it.

I wondered whether these students engaged in small-group work in ways that researchers have identified as productive behaviors. Webb and Mastergeorge (2003) indicate that students benefit from giving and receiving elaborated, conceptual explanations more than from getting answers or brief descriptions of procedures from peers in small groups. Also, students who seek help from peers benefit more if they ask for specific explanations rather than answers, if they persist in seeking help if they don't receive it the first time, and if they apply and use the help that they receive.

I thought that students would be more likely to engage in these more productive behaviors if they held values that aligned with the behaviors. Students who are more likely to give elaborated and conceptual explanations may be more likely to value working in small groups for understanding mathematics. Alternatively, students who are more likely to give procedural explanations or only share answers with peers when working in small groups may be more likely to value small-group work for finishing their work in class. Students who would be more likely to try to understand the thinking of their peers and support the learning of their group may be more likely to care about the needs of the group, not just their own needs.

To assess students' motivation to participate in small groups, I decided to ask students to talk about their experiences.

Designing Questions to Assess Students' Motivation

Next, I designed interview questions to elicit students' motives for engaging in small-group work. Although I planned to pose these questions to the students in an interview, I wanted to design questions that teachers could later use as prompts for student journals.

Identifying assumptions about students' motivation

Using my own assumptions about these students' motivations to participate in small groups, I speculated about reasons that students may give for why they did or did not participate. I related these assumptions to prior research on students' motivation, including the assumption that examining the values that an individual ascribes to an activity would be important.

Values can indicate a student's interest in a task, according to an expectancy–value perspective on motivation. Eccles (1983) described four task values: attainment value, intrinsic value, utility value, and cost. *Attainment value* refers to the relative importance that an individual assigns to succeeding at the task. *Intrinsic value* refers to personal enjoyment that someone derives from the task. *Utility value* refers to whether the task helps the individual achieve short- or long-term goals. *Cost* refers to negative consequences resulting from engaging in the task. Whether students perceive a particular activity to be worthwhile, and to what degree, will affect their engagement.

I assumed the following:

- Students may see some benefits in working together with peers.
- Adolescents like working in groups because the experience can feel less threatening than sharing their strategies or ideas about mathematics in front of the whole class during a large-group discussion.

Creating questions to examine assumptions

I created questions to pose directly to students about their motivation in relation to ideas from research and my assumptions. Concepts and results from research can be context dependent, so I wanted to assess whether they applied to these students in this classroom setting. Students' responses would allow me to revise my assumptions about their motivation either slightly or dramatically. I hoped that these questions would help me examine students' values for group work, whether students would speak about learning mathematics as an individual or collective experience because productive group work is collaborative, and the assumption in the *Principles and Standards* about whether they avoided participating because of being self-conscious (table 14.1).

Table 14.1
Questions designed to assess students' motivation to participate in small groups

Question	Alignment with task values
Why do you think your teacher has you work in groups in your math class?	Utility values (potential for achieving short- or long-term goals)
What do you like about working in groups in math class, if anything?	Intrinsic (personal enjoyment)
What do you dislike about working in groups, if anything?	Cost (negative consequences)
What would you say if your teacher stopped having you work in groups in your math class?	Attainment value (relative importance assigned to the task)

Table 14.1—*Continued*

Question	Alignment with task values
Does working in groups help you learn math? Why or why not?	Utility value (specifically for the goal of learning mathematics)
How often do you talk during small groups? Or do you mostly listen? Why?	Social dimension: students' roles in discourse

I intended to examine the social dimension of small-group work in this setting, because social goals affect students' learning (Urdan and Maehr 1995). Lampert, Rittenhouse, and Crumbaugh (1996) described a fifth-grade mathematics class in which some students expressed discomfort with being incorrect in front of their peers, whereas others appreciated the opportunity to hear how other students solved problems. To some students, being corrected during classroom discourse felt like a personal attack and affected how they felt about themselves and their classmates. These students "believed it was at least as important to maintain relationships as it was to argue mathematics" (Lampert, Rittenhouse, and Crumbaugh 1996, p. 756).

Listening to Students' Voices about Their Motivation to Participate in the Activity

After conducting interviews, I listened to the students' reasons for getting involved in small-group work (or not) and then elaborated on or revised my assumptions. Simultaneously, I was open to unexpected themes in the students' responses.

Sixth-Grade Students' Motivation to Participate in Small Groups: Their Task Values

The students who shared their motivations to participate in group work reported task values that indicated what they perceived to be the purpose of the group work (utility values), the relative importance of group work (attainment values), and the costs associated with group work. Students expressed utility values in response to interview questions designed to elicit them as well as questions designed to reveal intrinsic values.

Utility Value of Group Work

Utility values help us understand what goals these students attempted to achieve through engaging in small-group work. In response to interview questions, students said that they were interested in working in groups to learn mathematics, learn general social skills, and develop their mathematical autonomy.

Participate to learn mathematics

When students talked about how their small-group work helped them learn mathematics, they talked about the power of learning from peers. For instance, Clarissa said, "I like to work in groups so they can explain it to me." Iris mentioned, "[Classmates] don't tell me the answers, but, like, they try to give me, like, hints and stuff, so I can try to get it." Shawn said, "If you are stuck on a problem, they [peers] can help you out." Marisol described how she understood her peers' explanations more than her teacher's when she said, "Like, they [peers] can explain it better . . . like, let's say if [the teacher is] talking about something that I don't understand . . . I ask them [peers] first, and they explain it better." These students spoke about the experience in group work mostly as listening to peers' explanations or receiving help from their peers.

Less often, students said that working in small groups was an opportunity to help their peers (in contrast to how often students described wanting to work in groups to *receive* help from peers). Brian said, "I like working with groups most of the time because the group can help you out, and I can help the group out." Jasmine responded, "If they [peers] got it wrong, you can help them, like, get the right answer and stuff." For group work to be collaborative, students need to engage in receiving and giving help, and more students talked about receiving help than giving help.

Students described the content of what they were learning from their peers' explanations to be alternative solution strategies.

Interviewer: Why do you think your teacher has you work in groups?

Constance: So then you can back up your answers and get more multiple strategies, because not everyone thinks the same. And then if you get, if someone has something different, and you have something different than them, then you can work it out, and make it into a big strategy or something like that.

Constance said you can "make it into a big strategy," which suggested the teacher might have promoted looking for connections between the strategies. Students appeared to like the opportunity to hear additional solution strategies from peers.

Interviewer: And what do you like about working in groups, if anything?

Marisol: Like how we, you know, like [teacher] says, find multiple strategies? So we, like, we found, let's say we take . . . like, we are working by ourselves? Then after we are done we share with each other and we have two strategies, or something like that. . . .

Interviewer: Yeah. And you like that? [*Student nods.*] Why?

Marisol: Because we, like, we can learn from each other.

Marisol's interview excerpt revealed not only the utility value that she had for small-group work to be an opportunity for learning alternative solution strategies but also her intrinsic value for learning additional strategies. Other students also identified small-group work as an opportunity to learn alternative solution strategies and liked that purpose, such as Travis, who said, "I like to work with another person, and sometimes they might come up with a better strategy." When students described group work in mathematics class as useful to them, they often discussed an interest in receiving explanations about alternative strategies, but they did not say that they were looking only for answers from peers.

Participate to develop social skills

Students said that working with peers in small groups in their mathematics class was an opportunity to develop general social skills.

Interviewer: What do you like about working in a group in math class?

Bianca: You get to learn that other people learn different than you. And, like, you get to work together with people, and you get to learn other people's attitudes and stuff.

Interviewer: So why is that a good thing?

Bianca: Because you might not get along with somebody, but then you figure out they have the same as you, and then you start getting along with them and not fight so much.

Kaylee: Later on when we grow up we are going to have to work with people that maybe we might not like, [the teacher] tells us this, with people we might not like, but we have to work with them anyway. So [the teacher] kind of puts us with people maybe that we don't talk much with, and then we start getting used to working with people we don't really talk to anymore.

When students talked about participating to develop general social skills, these goals were longer term than their shorter-term goal of wanting to learn mathematics.

Participate to develop mathematical autonomy

Students reported that participation in small-group work allowed them to enact and develop their autonomy as mathematics learners.

Freddie: It's like, uh, instead of [the teacher] talking, we can decide, and [the teacher] doesn't tell us any answers, we have to figure it out on our own. So, it's, like, brain time. And you're just discovering the answer. It's fun . . . [the teacher] makes us discover it. We have to discover it, I have to discover it on my own. In order to learn it and get it in my system. My brain.

Marisol: So we can, like, think of it, like, we can get it by [ourselves]. Not just, like, [the teacher] being up there at the overhead just talking to us or something. Like, we can, like, brainstorm it and then us, like, saying stuff. Like, explaining the problems.

Students appreciated the opportunity to communicate and reason with each other about challenging mathematics.

Attainment Value of Group Work

Although students described how small-group work helped them achieve particular goals (such as learning mathematics), developing social goals, or achieving mathematical autonomy, students varied in how much importance they attributed to working in small groups. Some students preferred to work alone. Clarissa said, "Sometimes I like working by myself, 'cause . . . some people in my group, they don't like to work, like, as a team, and sometimes we usually end up working by ourselves anyway." When asked why he preferred to sit by himself, Travis said, "Because I get my work done better, and less distractions." These students talked about being able to complete their work better when they worked alone, and "better" implied more efficiently. Other students appeared to rely on group work heavily.

Interviewer: What if your teacher said, "Well, we're not going to work in small groups anymore; you have to work by yourself." What would you think?

Shawn: I would think, I probably can't do that, because I have to have some help.

When interpreting students' motivation, one must recognize that students may believe that small-group work can help them achieve a range of goals, but the opportunity to do so may be more or less important to particular students. Overall, most students in this sample had relatively strong attainment values for small-group work in mathematics class.

Cost of Small-Group Work

Some students also described the challenges of working in small groups in their mathematics classes. They described conflicts among group members and found themselves taking on the role of mediator to establish harmony.

Interviewer:	Okay, so do you think that you worked well together, or no?
Jeremy:	No, because they keep arguing. Like, [peer] will say something and [another peer] will get mad. So, it's hard to, like, cooperate, like.
Interviewer:	So what do you do when they're arguing?
Jeremy:	I try to get them to stop. They're arguing over stupid stuff.
Interviewer:	Is it about math, or no?
Jeremy:	No. It's just about stupid stuff. It's like, if [peer] is saying something, she'll get mad about it, and then they just start going back and forth, back and forth. So I just, like, forget about it. I mean, I try to get 'em to stop, but they don't.

Other students talked about one member of the group taking control and dominating the discussion.

Interviewer:	And what do you not like about it, if anything, when you work in groups?
Marisol:	When we discuss, like, when . . . let's say, like, we're trying to do group work, and then another person, he tries to take care of, like, the whole group . . . it's taking over our work. It's, like, he's taking over the whole group.

According to these students, at times the social dynamic in their groups detracted from learning mathematics. The students did not mention that participating in small groups was threatening, but small groups were not consistently safe for students to participate in, either, because of social conflicts.

Building on Students' Motivations: Creating Instructional Goals for Students

Knowledge about students' self-reported motivations could guide the creation of instructional goals for these students, as well as instructional strategies to support their achievement of these new instructional goals. When working to improve students' motivation and engagement, teachers must make the instructional goals broader than what can be accomplished in a single lesson. These goals are process-oriented goals that teachers should revisit over time.

Comparing students' voices, behaviors, and values that align with productive engagement in small-group work inspired new instructional goals. I considered whether students talked about their involvement in ways that aligned with these goals and created instructional goals to address potentially problematic issues that students mentioned. Students' voices also offered new insights about productive ways to engage in small-group work. Instructional goals addressed less productive motivations and extended and built on more productive motivations that students' talk revealed.

Table 14.2 describes the instructional goals for students' work in small groups that interviews with these students inspired. I purposely listed goals for instruction rather than descriptions of specific instructional interventions because their teachers could achieve these goals through a range of instructional strategies.

Table 14.2
Goals for instruction aligned with student interview findings

Student interview finding	Suggested instructional goal
Small-group work supports mathematics learning.	Students move toward monitoring whether *all* students in the group are learning. Consider learning of the group.
Small-group work supports learning to work with others.	Those students who have strong skills in this area (social skills, working well with others) can teach them to peers.
Small-group work supports development of mathematical autonomy.	Students who appreciate the opportunity to exercise their mathematical autonomy can help others have the opportunity to do so as well through active listening.
Small-group work decreases efficiency of task completion.	Students would benefit from developing a focus on mathematical understanding over task completion.

Table 14.2—*Continued*

Student interview finding	Suggested instructional goal
Some students are overly reliant on peers.	Students would benefit from attempting to solve problems themselves before listening to a peer's solution, and their group members would benefit from hearing their ideas.
Social conflicts occur during group work.	*All* students should take responsibility for monitoring the harmony of their interactions so that the same student is not always mediating conflict within a small group.
Peers can dominate the group work discussion.	Students would benefit from developing a value for opportunities to listen to a peer's solution in balance with sharing their own thinking.

Several themes appear across table 14.2's instructional goals: (*a*) moving toward an orientation of considering whether the *group* has learned, (*b*) monitoring a balance between individual needs and group needs, and (*c*) putting a priority on mathematical understanding over task completion. These instructional goals align well with the productive motivations that these students' talk revealed. Some students valued small-group work because they could learn general social skills. Explicitly teaching strategies for interacting productively in groups may appeal to these students. Other students valued the opportunity to learn multiple solution strategies. To build on this value, students could learn to propose increasingly elaborated and conceptual explanations of their strategies, which would support learning about mathematics from each other.

Conclusion

Teachers cannot observe students' behavior at all times during all instructional activities. Listening to how students talk about their motivation to participate in particular classroom activities can inform teachers regarding which students are and are not benefiting from current classroom practices and in what manner. This article illustrated assessing students' motivation to participate in small-group work and an application for using the data collected to enhance students' learning of mathematics through group work.

To assess students' motivation to participate in a particular activity, teachers could follow the model in this article: First, choose a focus activity and reflect on or read about productive behaviors for student engagement in that activity.

Next, review assumptions about students' motivation to participate in that activity, developed through experience or reading about students' motivation. These assumptions about students could influence the design of journal prompts or interview questions to determine whether students hold those motivations. The design of questions could be tailored to the specific activity, or they could be modifications of the general task value questions in table 14.1. Teachers should design questions that address the needs of their students.

To build on students' voices, teachers could then compare what their students said about participating to what teachers would want them to say about participating. Teachers could consider which motivational beliefs, goals, or values support productive engagement in a particular classroom activity. Then, teachers could design instructional strategies to support the development of these beliefs, goals, or values in connection with promoting productive behaviors. For students who hold productive motivational beliefs, goals, or values, teachers can explicitly validate students' experiences and articulate the connections that they see between what those students value or believe and the behaviors that these beliefs, goals, and values promote.

Teachers could also consider how their interactions with students can affect their students' engagement. The behaviors that students exhibit in small groups mirror the ways in which teachers interact with students in whole-class discussion, even after interventions have been implemented over an extended period to promote productive engagement in small groups (Webb, Nemer, and Ing 2006). If teachers would like their students to engage in effective help-giving practices, such as giving elaborated, conceptual explanations (instead of only sharing answers) and checking whether their peers understood their explanations, then teachers would need to model these sorts of explanations and checks for understanding during whole-class discussions. Also, if teachers want students to allow those who sought help to have a chance to apply the help that they received, teachers should not do all the cognitive work for students during the whole-class discussion.

The students' voices in this paper can inspire those teachers who want to get to know their students and can validate the efforts of those teachers who already work hard to understand their students' perspectives. Teachers are not necessarily effective at predicting what motivates their students (Middleton 1995), so teachers would benefit from developing strategies to learn about what motivates their students to engage productively in mathematics classroom activities. As we learn what motivates our students, we can challenge our assumptions about who is and who is not motivated. Through this sort of dedicated professional work of listening to students' voices, mathematics teachers can generate ideas for learning experiences that meet their students' needs more effectively.

REFERENCES

Dillon, James T. *Using Discussion in Classrooms*. Buckingham: Open University Press, 1994.

Eccles, Jacquelynne. "Expectancies, Values, and Academic Behaviors." In *Achievement and Achievement Motives*, edited by Janet T. Spence, pp. 75–146. San Francisco: W. H. Freeman, 1983.

Encyclopaedia Britannica. *Mathematics in Context*. Chicago: Encyclopaedia Britannica, 2006.

Hulleman, Chris S., Amanda M. Durik, Shaun B. Schweigert, and Judith M. Harackiewicz. "Task Values, Achievement Goals, and Interest: An Integrative Analysis." *Journal of Educational Psychology* 100 (May 2008): 398–416.

Lampert, Magdalene, Pamela Rittenhouse, and Carol Crumbaugh. "Agreeing to Disagree: Developing Sociable Mathematical Discourse." In *Handbook of Education and Human Development,* edited by David R. Olson and Nancy Torrance, pp. 731–64. Cambridge, Mass.: Blackwell, 1996.

Levin, Benjamin. "Putting Students at the Centre in Education Reform." *Journal of Educational Change* 1 (January 2000): 155–72.

Middleton, James A. "A Study of Intrinsic Motivation in the Mathematics Classroom: A Personal Constructs Approach." *Journal for Research in Mathematics Education* 26 (May 1995): 254–79.

National Council of Teachers of Mathematics (NCTM). *Principles and Standards for School Mathematics*. Reston, Va.: NCTM, 2000.

Urdan, Timothy C., and Martin L. Maehr. "Beyond a Two-Goal Theory of Motivation and Achievement: A Case for Social Goals." *Review of Educational Research* 65 (Fall 1995): 213–43.

Webb, Noreen M., and Ann M. Mastergeorge. "The Development of Students' Helping Behavior and Learning in Peer-Directed Small Groups." *Cognition and Instruction* 21 (February 2003): 361–428.

Webb, Noreen M., Kariane M. Nemer, and Marsha Ing. "Small Group Reflections: Parallels between Teacher Discourse and Student Behavior in Peer-Directed Groups." *Journal of the Learning Sciences* 15 (January 2006): 63–119.

Chapter 15

Motivating Mathematics Students with Manipulatives: Using Self-Determination Theory to Intrinsically Motivate Students

Brett D. Jones
Lida J. Uribe-Flórez
Jesse L. M. Wilkins

Mathematics teachers often discuss how manipulatives motivate their students. Although some research suggests that manipulatives in mathematics classrooms do indeed motivate students, the term *motivation* is often used generically to mean that students are having fun (e.g., Bolyard 2005; Moyer 2001; Reimer and Moyer 2005). Fewer researchers have framed their work with any major motivation theory to understand why students are motivated to use manipulatives. In this article, we consider how self-determination theory (SDT) can help explain why students are motivated to work on mathematics classroom activities involving manipulatives. We also discuss how educators can use this theory as a framework to become more intentional about using manipulatives to motivate students in mathematics classroom activities.

This paper consists of four sections. In the first section, we explain the difference between intrinsic and extrinsic motivation and why teachers should strive to intrinsically motivate students. Next, we briefly present SDT as a theory that can help explain why students are intrinsically motivated and the student needs that teachers should consider when trying to intrinsically motivate students. In the third section, we discuss definitions of manipulatives. Finally, we analyze two vignettes to demonstrate how different uses of manipulatives can have different effects on students' intrinsic motivation.

Intrinsic and Extrinsic Motivation

One of the most common and intuitive beliefs about motivation is that students are more likely to be motivated to choose and persist at an activity if they enjoy the activity and are interested in it. Researchers have defined such motivation as *intrinsic motivation* because students participate in an activity simply for its own sake (Schunk, Pintrich, and Meece 2008). That is, intrinsically motivated students enjoy or are interested in their work. Ryan and Deci (2000a) defined intrinsic motivation as "the inherent tendency to seek out novelty and challenges, to extend and exercise one's capacities, to explore, and to learn" (p. 70). In contrast, *extrinsic motivation* connects action to a separate outcome of the activity (Ryan and Deci 2000b). For example, students who do their homework to get a good grade but do not enjoy doing the homework are extrinsically motivated. Here the students are doing their homework to reach what they really want: a good grade. These students may also do their homework for other extrinsic reasons, such as avoiding punishment, pleasing parents, or proving to themselves that they can do it.

Intrinsic motivation results in high-quality learning and creativity (see Ryan and Deci 2000b for a discussion). Research has associated extrinsic motivation with negative outcomes, such as a higher likelihood of dropping out of school (Vallerand and Bissonnette 1992), student anxiety (Ryan and Connell 1989), lower creativity (Amabile 1983), and less flexible thinking (McGraw and McCullers 1979). Ormrod (2008) summarized research related to the advantages of intrinsic motivation over extrinsic motivation, noting that intrinsically motivated students are more likely than extrinsically motivated ones to

> pursue the task on their own initiative, without having to be prodded or cajoled; be cognitively engaged in the task (e.g., by keeping attention focused on it); undertake more challenging aspects of the task; strive for true understanding of the subject matter (e.g., by engaging in meaningful rather than rote learning); undergo conceptual change when such change is warranted; show creativity in performance; persist in the face of failure; experience pleasure, sometimes even exhilaration, in what they are doing; regularly evaluate their own progress, often using their own criteria; seek out additional opportunities to pursue the task; [and] achieve at high levels. (p. 454)

Teachers' striving to intrinsically motivate students makes sense because of intrinsic motivation's many positive outcomes and extrinsic motivation's negative outcomes, including generally leading to decreased intrinsic motivation (Deci, Koestner, and Ryan 1999).

Self-Determination Theory

SDT postulates that basic psychological needs are inherent in human life and assumes an organismic perspective in that individuals have needs from birth that evolve with development (Deci and Ryan 1985, 1991, 2008). Self-determined individuals can make choices and can manage the interaction between themselves and the environment to "engage in an activity with a full sense of wanting, choosing, and personal endorsement" (Deci 1992, p. 44).

SDT specifies three innate psychological needs that relate to intrinsically motivated processes: autonomy, competence, and relatedness (Deci and Ryan 1985). By supporting these three needs, teachers can increase students' intrinsic motivation (Ryan and Deci 2000a; Ryan and Deci 2002; Skinner and Edge 2002). Ryan and Deci claim that all three needs are important to an individual's being intrinsically motivated and that no one need is more important than the others.

Autonomy refers to an individual's control over his or her actions; autonomy can occur only when one's actions emanate from within oneself and are one's own (Deci and Ryan 1987). A student solving a problem by following a teacher's rules has less intrinsic motivation than a student solving the problem by choosing his or her own strategies and tools. *Competence* involves an individual's knowledge of how to accomplish an outcome and ability to perform the actions to achieve that outcome. Competence includes the desire to control results, as well as efficacy (Deci and Ryan 1991). For example, students who feel that they can complete an activity because they have the needed knowledge would be intrinsically motivated to work on the activity. *Relatedness* is a social psychological need to develop secure, satisfying relations with others. A student who feels related to a group will be motivated to engage in activities involving that group. (For a more detailed discussion of these three needs as they relate to SDT, see chapter 4 in this volume.)

In short, SDT states that individuals need to be in control of their actions (i.e., have their need for autonomy met), be good at what they do (i.e., have their need for competence met), and have secure and satisfying relationships with others (i.e., have their need for relatedness met). How do these three needs foster students' motivation? "Simply stated, social–contextual factors that afford people the opportunity to satisfy their needs for autonomy, competence, and relatedness will facilitate intrinsic motivation" (Deci and Ryan 1994, p. 7). The implication of SDT for educators is that teachers' social interactions with and activities for students can either support or diminish students' intrinsic motivation. Several researchers have studied teachers' social interactions with students by comparing the effects of "autonomy-supportive" teachers versus "controlling" teachers on students' intrinsic motivation (Deci and Ryan 1987). For example, students of teachers who were more autonomy oriented were more intrinsically motivated than students of teachers who were more control oriented (Deci et al. 1981). Students were also aware of the effects of their teachers' orientation: students' perceptions of their classroom climate correlated with their teachers' orientation. Students who perceived their teachers to be autonomy oriented reported higher levels of intrinsic motivation than did students who perceived their teachers to be control oriented (Ryan and Grolnick 1986; Vallerand, unpublished data 1991).

Definitions of Manipulatives

Teachers who use the word *manipulatives* to describe tools in their classroom may not be referring to the same things (Sherman and Richardson 1995). For some teachers, manipulatives are any concrete material that students can touch (e.g., boxes, calculators, chalkboard). For others, manipulatives are tools such as geoboards, strips of paper, or rulers. Determining what teachers mean when they attest to using concrete materials in their instructional practices can be difficult (Sherman and Richardson 1995).

Mathematics education researchers also differ in their definitions of manipulatives, which sometimes are contradictory. For Yeatts (1997), "manipulative materials are objects or things that appeal to several of the senses" (p. 7); thus, Yeatts would consider a video to be a manipulative because it appeals to one's vision and hearing. For others, manipulatives are objects that students can touch, move around and rearrange, or stack (Clement 2004; Hynes 1986; Kennedy 1986). For these researchers, books on the shelf are manipulatives even though students use the books only to read or solve problems. For others, manipulatives represent abstract mathematical ideas and are models of mathematical concepts that students can manipulate and that appeal to the senses (Durmus and Karakirik 2006; Moyer 2001; Suh 2005). Moyer (2001) defined manipulatives as "objects designed to represent explicitly and concretely mathematical ideas that are abstract" (p. 176). By this definition, manipulatives are commercial objects, such as Algeblocks, pattern blocks, or computer base/virtual manipulatives, because they represent mathematical ideas. However, this definition would exclude strings or rubber bands, even if students used them to understand a mathematical concept, because that purpose is not in such objects' design.

Another aspect of defining manipulatives involves the purpose for using the tools. Uttal, Scudder, and Deloache (1997) view manipulatives as objects designed specifically to help students learn mathematics. In contrast, McNeil and Jarvin (2007) define manipulatives as *any* object that helps students understand mathematics, even if not designed for that purpose. This latter definition could consider textbooks, pencils, chalkboards, strings, teacher-made pictures, and notebooks to be manipulatives because these objects could help students learn. Neither definition indicates how these objects called manipulatives help students learn mathematics. They could be called manipulatives because they require students to use their senses or because they allow students to make mathematical representations.

In the following discussion, we generally consider manipulatives to be any object that students or teachers can manipulate (e.g., can be stacked, moved around, folded, arranged) to represent mathematical ideas (e.g., concepts, relations). However, this definition of manipulatives would not limit using SDT to increase students' intrinsic motivation. We have chosen this definition simply to demonstrate how one can apply SDT in the following vignettes.

Fostering Intrinsic Motivation by Using Manipulatives: Two Vignettes

Figure 15.1 contains two vignettes that illustrate fairly typical use of manipulative-based activities in mathematics classrooms. Although students in both vignettes would feel self-determined to some degree, we believe that students in vignette 1 would feel more self-determined. As a result, students in vignette 1 would be more intrinsically motivated.

Vignette 1: The case of creative manipulation	Vignette 2: The case of the manipulator
Mr. Williams ran an orderly second-grade classroom in which his students were seated at desks arranged in tables, with four students to a table. Today, Mr. Williams was introducing subtraction of multidigit numbers by using base-ten blocks. Each table had a pile of base-ten blocks (small cubes, longs, and flats) in the center. To begin the lesson, he announced, "Today we are going to solve some mathematics problems. On the overhead projector you can see the problem that we're going to solve." He then read aloud the following problem: "Susie had $135 in her piggy bank. She wanted to buy a gift for her sister's birthday. Her sister really enjoyed the *Harry Potter* books. Susie found the whole *Harry Potter* series on sale for $47. If she bought the books, how much money would she have left?"	Mr. Beamer ran an orderly second-grade classroom. Students were usually on task and followed his every word closely. Individual desks were aligned neatly in rows facing the front of the classroom. Today, Mr. Beamer was using manipulatives to introduce students to subtracting multidigit numbers. Each student had a set of base-ten blocks (small cubes, longs, and flats), including a place-value mat with columns labeled 100s, 10s, and 1s. He also had an overhead mat to model the manipulations on the overhead projector. On the blackboard, he had written the following expression: $\begin{array}{r} 135 \\ \underline{-47} \end{array}$

Fig. 15.1. Vignettes illustrating students using manipulatives—*Continues*

Vignette 1: The case of creative manipulation	**Vignette 2: The case of the manipulator**
Students then used the base-ten blocks to solve the problem and discuss their solutions in their group. He told the children that once they had a solution, he wanted them to share it with the class. Mr. Williams circulated around the room to answer questions and ask probing questions of students having trouble. Some children initially worked alone with the base-ten blocks, whereas others immediately began talking and discussing the problem as a group. As students worked, several different solution strategies arose. Once each group had a solution to the problem, they presented it to the class. Some groups transferred their base-ten block solutions into pictures on the blackboard. Others chose to present their solution on the overhead. Still other groups presented a solution at their table while children gathered around to watch. Several different solution strategies were apparent. One group represented the $47 by using four longs and seven cubes, and they added blocks until they reached a total of $135, recognizing that the added quantity was the amount left over. Another group represented the $135 by using one flat, three longs, and five small cubes, and they worked through the trades that would allow them to subtract four longs and seven cubes, resulting in the difference. A third group modeled the $135 and $47 and, after trading one long for ten cubes, matched up seven cubes from the 47 with seven cubes from the 135 and laid them aside. Then they traded in the flat for ten longs and matched up the four longs from the 47 with four longs from the 135 and laid them aside, leaving them with eight longs and eight cubes (or $88). The class discussed the different solution strategies. At the end of class, the teacher asked the groups to create their own problem similar to the one that they had just solved and, again, solve it with the base-ten blocks.	He began by explaining: "Today we are going to learn how to subtract by using base-ten blocks." Then, he directed the students to represent 135 on their mats. Once he felt that every student had a chance to do so, he represented it on his overhead mat for students to check their work. Students who had represented the value incorrectly could change their answer. Next, he stated, "Now, we want to take away 47 from 135. Notice that we can't take 7 away from 5 [*pointing to the 7 and 5*], so we need to trade one of our three longs for 10 small cubes and place them in the 1s column." He modeled this process on the overhead mat and waited for students to complete the trade at their desks. He continued, "Now we can take away 7 cubes from the 15 cubes that we have in the 1s column, leaving us with 8 cubes." He waited to make sure that everyone had completed this step. "Next, we move to the 10s column, where we have 2 longs left. We can't take 4 away from two, so we have to trade in the flat for 10 longs and place them in the 10s column." Again, he waited for students to complete this step. "Once you have traded in the flat, you should have 12 longs in the 10s column. You can now take away 4 longs from the 12 longs, which leaves you with 8 longs." He modeled these steps on the overhead and waited for his students to complete the steps. He then asked the class, "How many are left?" to which some in the class responded, "eighty-eight." He put another subtraction problem on the board and asked the students to do the subtraction by using their base-ten blocks and mat. He moved around the room answering students' questions as they worked individually at their desks.

Fig. 15.1. Vignettes illustrating students using manipulatives—*Continued*

In the sections that follow, we compare the two vignettes to demonstrate how each meets or does not meet students' needs for autonomy, competence, and relatedness. An important point related to SDT is that students' level of intrinsic motivation depends on how they *perceive* that their needs are being met. In any one classroom, some students may believe that their needs are being met, whereas others may not. We do not mean our discussion to imply that all students will interpret any one classroom lesson the same way. Rather, our discussion offers an example of how teachers can think about how a lesson will most likely affect students' perceptions related to the three needs.

Autonomy

In vignette 1, students created their representations, chose to work in groups or individually (at least initially), represented the situation in a way that made sense to them, and presented their solution to the class in their own way. In all aspects, the teacher was autonomy supportive, and students were in control of their actions while working on the proposed task. Explaining and justifying answers, as students did during the presentation to the class, can be especially empowering (Borenson 1986). Moreover, Mr. Williams placed the problem in a specific context that students could probably relate to (i.e., buying something). If students saw this context as relevant to their lives, they would probably feel more autonomous because relevance also fosters feelings of autonomy (Assor, Kaplan, and Roth 2002).

In vignette 2, Mr. Beamer gave each student a set of base-ten blocks, and students could "manipulate" the objects at the beginning when he asked them to represent 135. Doing so allowed students to represent the number in their own way so that they had a sense of control and autonomy over their actions. However, that was the only chance that students had to be autonomous until the independent practice at the end of the lesson. For most of the lesson, the teacher was in control as he modeled each step of the process. Because students were simply mimicking his actions, they could not be as autonomous as the students in vignette 1.

Competence

Mr. Williams supported students' competence in several different ways in vignette 1. Because students had control of their actions and could use the manipulatives in their own way, they probably attributed their successes and failures to their own competence (and not to following the teacher's directions, as in Mr. Beamer's class in vignette 2). Of course, if they were unsuccessful, their perceived level of competence would probably decrease if they attributed their failures to lack of ability. Thus, it was incumbent on Mr. Williams to scaffold students' learning to lead them to success. Doing so required him to ensure that all students were successful. Because Mr. Beamer simply modeled the steps in the process during his lesson, he did not have this same responsibility until the end of his lesson when students were working on problems individually. We believe that, to feel self-determined in such classes, students must struggle in achieving their own success. We also believe that the teacher should help students as needed during this process; if the teacher does not, students can easily become frustrated and stop working on the task because they believe they do not have the abilities to be successful.

Working in groups allowed students in Mr. Williams's classroom not only to receive help from the teacher but also to scaffold one another's learning when they were unsure of a solution or made mistakes. When students shared their work with their partners, some might have felt more competent because they had the time to think and work on the activity individually before that point. While listening to a partner's explanation, students would have had an opportunity to evaluate their own thinking and increase the perception of competence in their own work. Sharing work with students in a group before sharing it with the whole class can be especially helpful for students who are unsure of their competence and hesitant to share with the class for fear of making a mistake and feeling stupid.

In vignette 2, Mr. Beamer started by allowing students to demonstrate their competence in representing 135 on their mat. If students had completed such an activity before, they would probably succeed and feel competent at this task. For the next part of the lesson, students relied on Mr. Beamer's actions to solve the tasks through how he modeled the steps. Students who successfully followed his steps probably felt competent. However, because students needed to follow and remember his actions to complete the tasks successfully, they relied on the teacher rather than on their own ability. Students may have believed that they could complete the tasks by following the teacher's directions, but they were not likely to be fully intrinsically motivated because they had no sense of control to go with the competence (Deci and Ryan 1991). Students feel competent when they succeed and believe that their success resulted from their own ability, over which they had control. Simply mimicking a teacher's actions might not give them the feeling that they could successfully complete such tasks on their own. Students may feel incompetent and anxious when teachers give them problems to solve on their own.

At the end of the lesson, Mr. Beamer assigned problems for students to complete individually and walked around the room answering questions. During this time, students had some autonomy, and if they succeeded, they may have felt competent at solving such problems. In this way, students may have been intrinsically motivated to engage in this activity, especially if they received any help that they needed from the teacher. This type of individual seatwork can give students clear feedback about their ability. Students who finish first are likely to believe that they are competent at these types of tasks compared with their classmates. However, students who are among the last to finish will probably believe that they are not very good at such tasks compared with their classmates. As a result, such individual seatwork assignments can boost some students' perceptions of competence and lower others.

We believe that teachers in both vignettes would have some success in teaching students with manipulatives. Students working with manipulatives can feel that manipulating objects and making meaning of those actions is easier than manipulating symbols such as numbers or letters. Before actually using manipulatives in a mathematics activity, students should have had a chance to become familiar with them through other activities. Doing so would probably help them feel more competent during the actual lesson. Also, because manipulatives offer a concrete representation of mathematical concepts, these tools may help students feel that the content is less abstract and more adjusted to their skills and capabilities. Consequently, students may feel that the activity is more readily accomplishable than simply manipulating abstract symbols such as numbers and letters.

Relatedness

The two vignettes present differences in supporting students' need for relatedness. First, the layout of the classrooms is different. Mr. Williams supplied opportunities for students to interact with one another with a setup for small-group work, and students worked together for much of the lesson. Even students who chose to work individually during the first part of the lesson could engage in group discussions that could help them create close relations with their peers and feel part of the group. Simply having students work together does not necessarily meet their need for relatedness. Sometimes the opposite effect occurs, as when students dislike other students in their group or some students isolate or disparage others. Meeting students' relatedness needs requires developing caring relationships that are secure and satisfying. Therefore, if Mr. Williams does not help students work with one another successfully, his lesson may not foster relatedness and ultimately will prove to be less intrinsically motivating to students.

Student interaction in Mr. Beamer's classroom was limited because students sat at desks arranged in rows facing the teacher. Thus, students' interactions probably would not foster a sense of belonging to the group. Because the teacher talked much of the time and student discussion was nonexistent, opportunities for interaction among students and between the students and the teacher were limited. The last part of Mr. Beamer's lesson, in which the students worked independently, fostered no relatedness, except maybe a caring relationship between the teacher and the student. Students may have felt that the teacher cared enough to come around and help them. However, this type of independent seatwork might have enhanced competitiveness because discerning who finished their assignments first and last was easy. Such competitiveness can diminish the sense of community in the classroom and increase the importance of individual achievement.

Vignette Summary

On the whole, we believe that the instruction in vignette 1 is more likely than that in vignette 2 to foster students' self-determination and intrinsic motivation. We make this assertion on the assumption that teachers instruct in a manner that is consistent with our prior discussion and that enhances students' autonomy, competence, and relatedness. Instruction not enhancing these three needs would probably not intrinsically motivate students. In this way, motivating students intrinsically is similar to implementing any other type of instruction in that the details matter and can dramatically affect the overall success of the lesson. Suppose that the students in vignette 1 did not know how to work well with other students in their group. This lesson would frustrate students, and they would neither be intrinsically motivated nor learn the concepts. However, with correct implementation, the instruction in vignette 1 would be a powerful way to meet students' three needs. In comparison, although vignette 2's instruction probably has fewer opportunities for problems to arise, this implementation offers less autonomy support, making students less likely to enjoy the lesson.

Conclusion

On the basis of SDT, we conclude that not all uses of manipulatives in the mathematics classroom intrinsically motivate students equally. *How*, not whether, teachers use manipulatives matters most in intrinsically motivating students. Teachers can foster students' intrinsic motivation when they use manipulatives in ways that support students' autonomy, develop their competence, and allow them to experience relatedness with the teacher and other students. We hope that our analysis of the two vignettes has demonstrated that fostering students' intrinsic motivation through meeting these three needs is a complex, yet achievable, goal. We acknowledge that knowing whether any one lesson will meet every student's needs is impossible, but by considering these needs, teachers are more likely to create and implement lessons that intrinsically motivate students.

REFERENCES

Amabile, Teresa M. *The Social Psychology of Creativity*. New York: Springer, 1983.

Assor, Avi, Haya Kaplan, and Guy Roth. "Choice Is Good, but Relevance Is Excellent: Autonomy-Enhancing and Suppressing Teacher Behaviours Predicting Students' Engagement in Schoolwork." *British Journal of Educational Psychology* 72 (June 2002): 261–78.

Borenson, Henry. "Teaching Students to Think in Mathematics and to Make Conjectures." In *Teaching Mathematics: Strategies That Work K–12,* edited by Mark Driscoll and Jere Confrey, pp. 63–70. Portsmouth, N.H.: Heinemann, 1986.

Bolyard, Johnna J. "A Comparison of the Impact of Two Virtual Manipulatives on Student Achievement and Conceptual Understanding of Integer Addition and Subtraction." Ph.D. diss., George Mason University, 2005.

Clement, Lisa. "A Model for Understanding, Using, and Connecting Representations." *Teaching Children Mathematics* 11 (September 2004): 97–102.

Deci, Edward L. "The Relation of Interest to the Motivation of Behavior: A Self-Determination Theory Perspective." In *The Role of Interest in Learning and Development,* edited by K. Ann Renninger, Suzanne Hidi, and Andreas Krapp, pp. 43–70. Hillsdale, N.J.: Erlbaum, 1992.

Deci, Edward L., Richard Koestner, and Richard M. Ryan. "A Meta-Analytic Review of Experiments Examining the Effects of Extrinsic Rewards on Intrinsic Motivation." *Psychological Bulletin* 125 (November 1999): 627–68.

Deci, Edward L., and Richard M. Ryan. *Intrinsic Motivation and Self-Determination in Human Behavior.* New York: Plenum, 1985.

———. "The Support of Autonomy and the Control of Behavior." *Journal of Personality and Social Psychology* 53 (December 1987): 1024–37.

———. "A Motivational Approach to Self: Integration in Personality." In *Perspectives on Motivation,* edited by Richard Dienstbier, pp. 237–88. Vol. 38, Nebraska Symposium on Motivation. Lincoln, Neb.: University of Nebraska Press, 1991.

———. "Promoting Self-Determined Education." *Scandinavian Journal of Educational Research* 38, no. 1 (1994): 3–14.

———. "Facilitating Optimal Motivation and Psychological Well-Being across Life's Domains." *Canadian Psychology* 49, no. 1 (February 2008): 14–23.

Deci, Edward L., Alan J. Schwartz, Louise Sheinman, and Richard M. Ryan. "An Instrument to Assess Adults' Orientation toward Control versus Autonomy with Children: Reflections on Intrinsic Motivation and Perceived Competence." *Journal of Educational Psychology* 73 (October 1981): 642–50.

Durmus, Sonor, and Erol Karakirik. "Virtual Manipulatives in Mathematics Education: A Theoretical Framework." *Turkish Online Journal of Educational Technology* 5 (January 2006): 117–23.

Hynes, Michael C. "Selection Criteria." *Arithmetic Teacher* 33 (February 1986): 11–13.

Kennedy, Leonard M. "A Rationale." *Arithmetic Teacher* 33 (February 1986): 6–7.

McGraw, Kenneth O., and John C. McCullers. "Evidence of a Detrimental Effect of Extrinsic Incentives on Breaking a Mental Set." *Journal of Experimental Social Psychology* 15 (May 1979): 285–94.

McNeil, Nicole M., and Linda Jarvin. "When Theories Don't Add Up: Disentangling the Manipulative Debate." *Theory into Practice* 46 (September 2007): 309–16.

Moyer, Patricia. "Are We Having Fun Yet? How Teachers Use Manipulatives to Teach Mathematics." *Educational Studies in Mathematics* 47, no. 2 (2001): 175–97.

Ormrod, Jeanne L. *Human Learning*. 5th ed. Columbus, Ohio: Merrill Prentice Hall, 2008.

Reimer, Kelly, and Patricia S. Moyer. "Third-Graders Learn about Fractions Using Virtual Manipulatives: A Classroom Study." *Journal of Computers in Mathematics and Science Teaching* 24, no. 1 (2005): 5–25.

Ryan, Richard M., and James P. Connell. "Perceived Locus of Causality and Internalization: Examining Reasons for Acting in Two Domains." *Journal of Personality and Social Psychology* 57 (November 1989): 749–61.

Ryan, Richard M., and Edward L. Deci. "Self-Determination Theory and the Facilitation of Intrinsic Motivation, Social Development, and Well-Being." *American Psychologist* 55 (January 2000a): 68–78.

———. "Intrinsic and Extrinsic Motivations: Classic Definitions and New Directions." *Contemporary Educational Psychology* 25 (January 2000b): 54–67.

———. "An Overview of Self-Determination Theory: An Organismic-Dialectical Perspective." In *Handbook of Self-Determination Research*, edited by Edward L. Deci and Richard M. Ryan, pp. 3–33. Rochester, N.Y.: University of Rochester Press, 2002.

Ryan, Richard M., and Wendy S. Grolnick. "Origins and Pawns in the Classroom: Self-Report and Projective Assessment of Individual Differences in Children's Perceptions." *Journal of Personality and Social Psychology* 50 (March 1986): 550–58.

Schunk, Dale H., Paul R. Pintrich, and Judith L. Meece. *Motivation in Education: Theory, Research, and Applications*. Upper Saddle River, N.J.: Merrill Prentice Hall, 2008.

Sherman, Helene, and Lloyd Richardson. "Elementary School Teachers' Beliefs and Practices Related to Teaching Mathematics with Manipulatives." *Educational Research Quarterly* 18 (June 1995): 27–37.

Skinner, Ellen, and Kathleen Edge. "Self-Determination, Coping, and Development." In *Handbook of Self-Determination Research,* edited by Edward L. Deci and Richard M. Ryan, pp. 297–337. Rochester, N.Y.: University of Rochester Press, 2002.

Suh, Jennifer. "Third Graders' Mathematics Achievement and Representation Preference Using Virtual and Physical Manipulatives for Adding Fractions and Balancing Equations." Ph.D. diss., George Mason University, 2005.

Uttal, David H., Kathryn V. Scudder, and Judy S. DeLoache. "Manipulatives as Symbols: A New Perspective on the Use of Concrete Objects to Teach Mathematics." *Journal of Applied Developmental Psychology* 18 (January–March 1997): 37–54.

Vallerand, Robert J., and Robert Bissonnette. "Intrinsic, Extrinsic, and Amotivational Styles as Predictors of Behavior: A Prospective Study." *Journal of Personality* 60 (September 1992): 599–620.

Yeatts, Karol. *Manipulatives: Motivating Mathematics*. Miami: Dade Public Education Fund, 1997. ERIC Document Reproduction no. ED355097.

Chapter 16

Using Movies and Television Shows as a Mathematics Motivator

Elana Reiser

In recent years, popular culture has joined, perhaps accidentally, the effort to improve the U.S. educational system. More and more television shows and movies have incorporated mathematical topics into their scripts and, by doing so, offer the idea that mathematics is both trendy and functional. Given that students are more interested in learning when the topic is relevant to them, popular culture offers a necessary means for student engagement (Brophy 2004). This article will explore contemporary movies and television shows that include one or more mathematical elements and will suggest how mathematics teachers might use such media to boost student motivation.

Garnering student interest is a long-documented battle. Deci and Ryan (1987) found a correlation between intrinsic motivation and self-reports of interest. Culturally, students of all ages have a significant interest in movies and television; Beckmann, Thompson, and Austin (2004) describe how that interest can motivate students to learn mathematics: "If we can draw the mathematics out of the movies and literature that students find engaging, then we can make mathematics more meaningful and interesting to students" (p. 261).

We will look at several examples of mathematics in movies and television through the lens of the National Council of Teachers of Mathematics (2000) five suggested Content Standards: Number and Operations, Algebra, Geometry, Measurement, and Data Analysis and Probability. I will give examples from movies or television shows for each Content Standard. A suggestion for why you might want to show the clip in your classroom and how you can use it to initiate a discussion or teach or reinforce a lesson will follow each example.

Linking Movies and TV Shows to Number and Operations

We can link the movie *Matilda* to the Number and Operations Standard. The main character, a young girl named Matilda, has a gift for quickly adding and multiplying numbers in her head, which she exhibits in several scenes. In one such scene, Matilda's dad asks his son to document the prices for which he has sold several used cars and to add them with pen and paper. Matilda can sum the numbers in her head faster than her brother can by hand. When using this movie, you might have students think about how Matilda can add numbers in her head quickly. After this discussion you might examine other interesting mental strategies such as this one for multiplying by 11. One way to mentally multiply 1,784 × 11 is to *first do 1,784 × 10 = 17,840 and then sum 17,840 + 1,784 = 19,624.* Another strategy for multiplying by 11 is to add two consecutive digits from left to right along with completing the required regrouping. With this approach, one finds 1,784 × 11 by placing 4 in the ones place; adding 8 and 4, recording the 2 in the tens place; adding 8, 7, and the 1 from the prior regrouping of 8 and 4, recording the 6 in the hundreds place; adding 1, 7, and the 1 from prior regrouping of 8, 7, and 1, recording the 9 in the thousands place; and placing the 1 in the ten thousands place. Depending on the grade level for which you are using this, discussing and justifying why this method works might be appropriate. Many resources exist for finding additional mental mathematics. You can find some on the Mental Math Power Zone (www.themathlab.com/natural/mental%20math%20tricks/powerzone.htm), and 30 Fast Mental Math Tricks (www.glad2teach.co.uk/fast_maths_calculation_tricks.htm) has videos.

The television show *Futurama* features another example of number and operations. The writers chose 1729 for inclusion in a scene because it is the smallest integer that can be expressed in two different ways as the sum of two cubes (of positive integers): $1^3 + 12^3 = 1 + 1728$ and $9^3 + 10^3 = 729 + 1000$. From that point forward, the writers exclusively chose sums of two cubes whenever referring to numbers. Students may be interested in the conversation between two famous mathematicians, G. H. Hardy and Srinivasa Ramanujan, regarding this number (mathworld.wolfram.com/Hardy-RamanujanNumber.html). After students see 1729 in some episodes, have them come up with more examples of these types of numbers. During lunch or study hall, students can watch episodes of *Futurama* and look for such numbers. Also, have students think about other special numbers, why one might use them, and whether they can invent their own.

If students find the work with cubes interesting, you might ask them to explore relationships that they can notice in the following explorations. You might ask what other patterns they can generate.

1. What patterns do you notice?

   a. $1^3 + 2^3 =$

   b. $1^3 + 2^3 + 3^3 =$

   c. $1 + 2^3 + 3^3 + 4^3 =$

2. What patterns do you notice?

   a. 23 + 33 =

   b. 33 + 43 =

   c. 43 + 53 =

Another example that fits into this standard is from *The Da Vinci Code*. French police ask the main character, Robert Langdon, to look at the crime scene where the curator of the Louvre has died. A secret message written in invisible ink is near him. It reads as follows:

13 – 3 – 2 – 21 – 1 – 1 – 8 – 5
O, Draconian devil!
Oh, lame saint!

Although the message at first appears cryptic, Langdon realizes that the numbers are a mixed-up version of the Fibonacci numbers, 1, 1, 2, 3, 5, 8, 13, 21, 34, ... He rearranges the letters and gets "Leonardo Da Vinci the Mona Lisa." Teachers can use this clip to introduce cryptography, the science of analyzing and deciphering codes. Younger students can use a simple coding system such as the Caesar cipher. This system allows for coding and decoding messages by assigning each letter to the one that is three spots away (www.simonsingh.net/The_Black_Chamber/caesar.html). Older students can use more complicated ciphers such as RSA encryption, a method that involves prime numbers and modular arithmetic (mathworld.wolfram.com/RSAEncryption.html). Myerscough et al. (1996) and Gorini (1996) give cryptography examples of all different levels. After students see Langdon solve the riddle, they may become more interested in trying one themselves. Teachers might have students work in pairs. One student chooses a secret message to encode by using the cipher, and the other student decodes the message. After trying this in my classroom, one student told me that he taught the method to his friend so that they could send secret messages to one another.

Teachers can also use this clip to introduce the Fibonacci numbers. Depending on the grade, teachers can find elementary-level (mathforum.org/library/topics/golden ratio) or other age-appropriate activities. The Fibonacci sequence is defined recursively, with each element after the first two defined based on previous elements. The ratio of the consecutive terms, 1/1, 2/1, 3/2, 5/3, 8/5, 13/8, 21/13, . . . converge to a limiting value called the golden ratio, Φ. Teachers can also discuss the idea of recursion or introduce talk about limits. Teachers may ask students to pick any two starting numbers and build a sequence in a manner similar to the Fibonacci sequence. For example, if we start with 5 and 12, we would generate the sequence 5, 12, 17, 29, 46, 75, 121, 196, ... The teacher can have students investigate the values formed by the ratios of consecutive terms of this or other sequences similarly generated.

Linking Movies and TV Shows to Algebra

Tom Hanks's character in *Big* uses basketball as an analogy to help his friend's son learn algebra: if you score 10 points in the first quarter, how many points would you probably score in a game? Another example is in the movie *Little Big League*, where a 12-year-old baseball team owner makes his players work on this algebra problem: if it takes one man 3 hours to paint a house and a second man needs 5 hours, how long will it take both of them working together? Using such clips in the classroom allows students to see that they are learning material embedded in popular culture.

A book by Chappell and Thompson (2009) gives other examples using algebra. These examples are a little different because they use a movie as the basis for an activity. One movie that the book uses is *Akeelah and the Bee*. Akeelah is studying to be in a spelling bee and must learn to spell many new words. An activity sheet asks students to think about the number of words that she has to learn paired with how much time she has to learn them. Students use algebra to come up with equations relating these two values.

Another algebra example in Chappell and Thompson's book uses the movie *The Pursuit of Happyness*. In this movie the main character, Chris, sells bone density scanner machines, but the original price he charges is unknown. The activity sheet has students use algebra with the possible prices he charges at various parts of the movie. One question asks students to write a piecewise function with the amount that Chris earned as one variable and the number of machines that he sold as another variable.

Linking Movies and TV Shows to Geometry

A situation involving geometry appears in the movie *Death and the Compass*, based on the short story by Jorge Luis Borges. The movie features three murders occurring at locations that form an equilateral triangle (fig. 16.1). Contrary to his police force cronies, the inspector predicts that a fourth murder will occur, at a position that would result in a rhombus if the four locations were joined. Three possible locations will form a rhombus (fig. 16.2).

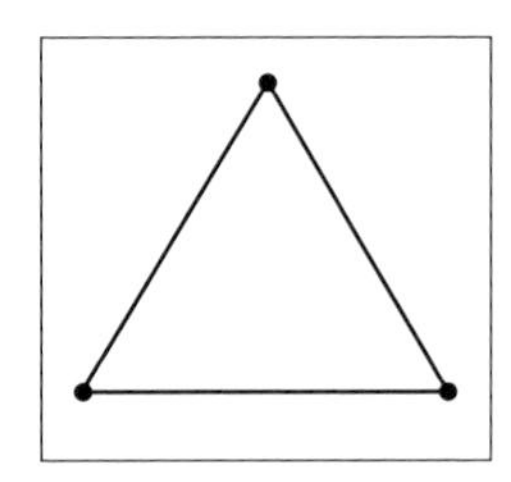

Fig. 16.1. Equilateral triangle

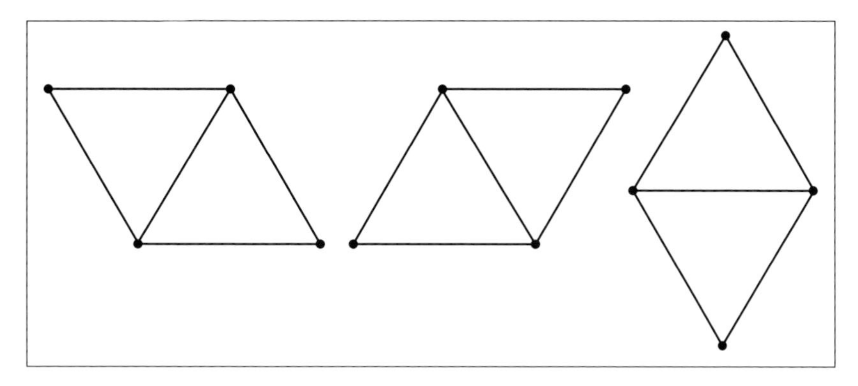

Fig. 16.2. Rhombuses formed from an equilateral triangle

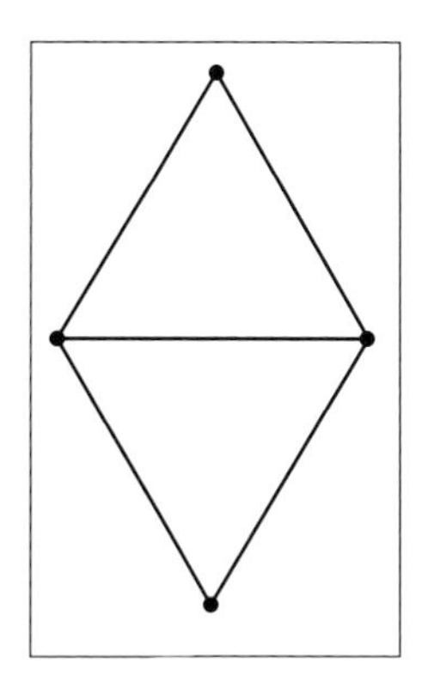

Fig. 16.3. The rhombus that accurately predicts the fourth murder

The clues supplied suggested that since the first three crimes occurred in the north, east, and west, the remaining crime would occur in the south. Therefore, the inspector knew that the rhombus in figure 16.3 would be the correct one.

Chuck, the protagonist in *Cast Away,* is another character who illustrates using geometry as well as measurement. Chuck figures out how many square miles rescuers would have to search to rescue him after his plane crashes. He first estimates the distance from where the plane was supposed to be and where it ended up. He then multiplied the square of this value by π to find the area of the circle that needs to be searched.

After showing students this clip, teachers can encourage them to consider why using the circle that Chuck selected made sense.

Linking Movies and TV Shows to Measurement

Though examples of the Measurement Standard are rare, *Die Hard: With A Vengeance* provides a good one. Bruce Willis and Samuel L. Jackson's characters hunt a criminal by solving riddles. In one scene, they use a five-gallon jug and a three-gallon jug to measure exactly four gallons of water. During class, you can first show the clip where the criminal explains the riddle and sets forth the rules. After viewing, have students work out the solution and discuss their findings. Show the rest of the clip where the characters solve the riddle.

I used this example with a group of nonmathematics majors on the first day of a college class that I taught. The topic was problem solving, and normally I would have explained Polya's methods and then had students work in groups to solve various word problems. When I showed them this movie clip, the students got excited about finding the solution. Once they solved the problem, I showed the clip where the characters in the movie solved it. The students were gratified knowing that through their own hard work they could solve a riddle worthy enough to be in a movie. One student even said, "I have seen this movie a bunch of times, but I never stopped to think about how they solved that riddle." It seemed that using the movie clip for the first example excited my students about the topic and that they enjoyed solving the other problems—a welcome motivator, especially since these students took the course only to fulfill their mathematics requirement.

Linking Movies and TV Shows to Data Analysis and Probability

The Da Vinci Code also presents an example of the Data Analysis and Probability Standard. Teachers can use this clip best when students are first learning permutations. One character, Sophie, describes how many permutations are possible on a *cryptex*, a term that Dan Brown coined for the novel. A cryptex is a device that holds a scroll with a secret message. It is similar to a combination lock with five wheels that each contain the twenty-six letters of the alphabet. Knowing that it was a five-letter code, a teacher might have students think about how Sophie could solve the problem, before sharing that she computed the answer, 11,881,376, by multiplying $26 \times 26 \times 26 \times 26 \times 26$. A teacher can then give similar problems to stretch the exercise further.

The television show *Monk* offers another example of using probability. In one episode, two female victims have the same first and last names. The characters want to know how likely a coincidence it is. One detective tries to explain why this finding isn't as unlikely as one might think, but he botches his explanation after getting flustered. You can have students think about the detective's answer and how they might calculate the probability. According to the U.S. Census Bureau (1990), 88,799 last names and 4,275 female first names are possible. With this information, have students calculate the probability of two females having the same first and last names. [Answers will be estimates only because the census information (*a*) is rough and (*b*) doesn't take into account that certain first and last names more commonly go together.]

Teachers can lead this discussion to other types of probability problems such as the Monty Hall problem from television. The Monty Hall problem is as follows: "On a game show, there are three doors. Behind one is a car; the others have goats. The contestant picks a door, and the host, who knows what's behind each, opens one of the doors the contestant did not pick. Behind it is a goat. The host offers the contestant the option to switch the current choice for the remaining unopened door. Should the contestant switch?"

In the movie *21*, a mathematics professor posed the Monty Hall problem to his students to find out who was smart so that he could recruit them to count cards in blackjack. The question also appeared in *Parade* magazine's "Ask Marilyn" column, which features Marilyn vos Savant, the *Guinness Book of World Records* titleholder for highest IQ. A *New York Times* article (Tierney 1991) documents her experience. When vos Savant answered that the contestant should switch doors, she received a flurry of letters from all over the globe, many from mathematics professors, saying that she was wrong. But she was right, which shows students that probability is sometimes counterintuitive, even to professors, and that problems from mathematics can be of interest to the general public.

Summary

The preceding are a few examples that teachers might use if they want to bring mathematics from movies or television to their students. However, many others also exist. The "Math in the Movies" Web page (world.std.com/~reinhold/mathmovies.html) is an excellent resource. Teachers can also find other examples in published articles. For example, Takis (1999) explains how she used the movie *Titanic* to teach statistics; Wood (1992) discusses different games from *The Price is Right* to teach probability and problem solving; Appelbaum (1995) uses *Wheel of Fortune* to teach problem solving; Beckmann, Thompson, and Austin (2004) use movies such as *The Perfect Storm* and *Harry Potter* to teach proportional reasoning; and Chappell and Thompson (2009) draw from movies such as *The Pursuit of Happyness* in a book aimed at tying mathematics into popular culture. Using mathematics to solve crimes is a popular concept in television shows such as *Law & Order* and *CSI*. Teachers can find a plethora of activities to complement episodes of *NUMB3RS* on Texas Instruments's activity exchange Web site (education.ti.com/educationportal). Teachers can also make up a fake crime for students to solve or have students develop their own and switch with a classmate for solving.

Another great option is to ask students to be on the lookout for mathematics in the movies and television shows that they watch. When students see such an example, they will be more likely to discuss it with their friends and family and, in turn, better understand it themselves. Incorporating these media clips into lessons will not require much extra time. Most of the clips in this chapter are less than five minutes long, and some are even shorter than one minute.

The primary purpose of the preceding examples is to show the potential for entertainment as a means of motivation. The examples show that mathematics can be relevant to a student's life by highlighting math's pervasive presence in the entertainment that they frequently watch.

If our challenge as teachers is, as Deitte and Howe (2003) suggest, motivating students to feel responsible for their educational development, we need to understand what drives them. "This goal," Deitte and Howe continue, "is most effectively accomplished if students convince themselves that mathematics is interesting and useful" (p. 278). Equating mathematics with popular culture serves that precise purpose.

REFERENCES

Appelbaum, Peter M. *Popular Culture, Educational Discourse, and Mathematics.* Albany, N.Y.: State University of New York Press, 1995.

Beckmann, Charlene E., Denisse R. Thompson, and Richard A. Austin. "Exploring Proportional Reasoning through Movies and Literature." *Mathematics Teaching in the Middle School* 9 (January 2004): 256–62.

Brophy, Jere. *Motivating Students to Learn.* Mahwah, N.J.: Lawrence Erlbaum, 2004.

Chappell, Michaele F., and Denisse R. Thompson. *Activities to Engage Middle School Students through Film, Literature, and the Internet.* Portsmouth, N.H.: Heinemann, 2009.

Deci, Edward L., and Richard M. Ryan. "The Support of Autonomy and the Control of Behavior." *Journal of Personality and Social Psychology* 53 (December 1987): 1024–37.

Deitte, Jennifer M., and Michael Howe. "Motivating Students to Study Mathematics." *Mathematics Teacher* 96 (April 2003): 278–80.

Gorini, Catherine A. "Using Clock Arithmetic to Send Secret Messages." *Mathematics Teacher* 89 (February 1996): 100–4.

Myerscough, Don, Don Ploger, Lynn McCarthy, Hallie Hopper, and Vicki G. Fegers. "Cryptography: Cracking Codes." *Mathematics Teacher* 89 (December 1996): 743–50.

National Council of Teachers of Mathematics (NCTM). *Principles and Standards for School Mathematics.* Reston, Va.: NCTM, 2000.

Takis, Sandra L. "Titanic: A Statistical Exploration." *Mathematics Teacher* 92 (November 1999): 660–64.

Tierney, J. "Behind Monty Hall's Doors: Puzzle, Debate, and Answer?" *New York Times* 21 July 1991, p. 1.

U.S. Census Bureau. "Frequently Occurring First Names and Surnames From the 1990 Census." 1990. www.census.gov/genealogy/names (accessed October 19, 2009).

Wood, Eric. "Probability, Problem Solving, and 'The Price Is Right.'" *Mathematics Teacher* 85 (February 1992): 103–9.

Part IV
Professional Development Support

PART IV of this yearbook is titled "Professional Development Support" and features five articles. This section explores the role that professional development activities can and do play in affecting the motivation and disposition of teachers and students toward the study of mathematics.

"Bringing Cool into School" takes an interesting perspective on what teachers can do to motivate students. The author skillfully points out that students are often motivated to persist at challenging tasks in other domains—but not always in mathematics. He attributes this to the notion that both the public and our students perceive mathematics as decidedly "uncool." He offers specific suggestions for what a teacher might do to make math "cool."

"Venturing into Storybooks: Motivating Teachers to Promote Mathematical Problem Solving" makes the point that an important part of professional development involves motivating teachers to look for interesting ways to engage their students with mathematics. Building on primary teachers' love of sharing storybooks with their students, the authors help to illuminate how teachers can travel through a set of phases that transform teachers in their motivation to encourage students to use their imaginations to solve mathematical problems. The authors point out, however, that motivation does require thoughtful guidance and modeling.

The authors of "TARGETTS: A Tool for Teachers to Promote Adaptive Motivation in Any Mathematics Lesson" share the professional-development potential of a research-based tool for teachers to analyze their own classroom practice in the context of improving adaptive motivation. They have found through such a professional experience how teachers' ideas of what motivates students can change in interesting ways. One of the most compelling aspects of the chapter is the recognition that making instructional practices explicit for teachers to consider is often the key to teacher change.

"Developing Prospective Teachers' Productive Disposition toward Fraction Operations" asserts that developing teacher candidates' willingness to persevere and to attend to sense making in mathematics is vital. The authors make the case that using contextual and visual representations of mathematics is particularly effective at developing this willingness. The authors share their identification of the phases that preservice teachers go through in this quest by focusing on fraction operations, a topic in which teachers often lack confidence.

"Transforming Mathematics Teachers' Attitudes and Practices through Intensive Professional Development" narrates a long-term research project partnering universities and school systems in a mission to change teachers' disposition toward using reform-based mathematics teaching. Seeing the evidence that teachers working in school-based cohorts, supporting each other, and receiving support from professional experts can significantly change their pedagogical perspectives is fascinating.

While reading the chapters in this section, you might consider some of these questions:

- Do you think that teachers can learn to be "cool" or show passion in their instruction, or do they already need to have a disposition toward being overtly enthusiastic?
- Should passion and enthusiasm in teaching mathematics be a requirement for entry to the profession?
- Can teachers above the primary level also be motivated to stretch students' interest in solving mathematical problems through using storybooks? How?
- Is identifying the phases that teachers, whether preservice or in-service, go through as their practice is transformed important? To whom might doing so be important?
- Is it true that motivation techniques need to be linked to building students' conceptual understanding of mathematics, or can flashier techniques also work? Have you ever seen them work, and do trade-offs exist?
- If working in a cohort situation supports teacher motivation to improve their mathematical instruction, what happens to a teacher teaching in an environment where others are not open to these changes? What can a teacher teaching in an isolated area do to develop a cohort of colleagues?

M. Lynn Breyfogle and Marian Small

Chapter 17

Bringing Cool into School

Keith Devlin

"OH, NO, not again," sighed Sally when she looked up from the chopping board and glanced out the kitchen window.

Dropping the knife into the pile of half-diced vegetables, she grabbed a towel and started to dry her hands as she headed out through the open doorway.

"That's far enough. Stop there, young man," she commanded her fourteen-year-old son. "We're going to clean the dirt and blood off you before you come into my nice, clean kitchen." Sean grinned back at her. "And you've chipped a tooth," Sally added.

She took the skateboard from his hand and quickly looked him over from head to toe. "Where does it hurt?"

"Nowhere really, apart from the grazes." Sean cast his eyes over his blood-caked right elbow and forearm. "And my knee," he added, seeming to see for the first time the swelling that was already turning bright purple.

"What was it this time?"

Sean ran his hand through his tousled, and decidedly dirty, hair. "A kick flip."

"That's where . . ." She struggled to remember the explanation that he had almost certainly given her a dozen times before.

Sean rescued her: "It's like an Ollie, but you kick the board over while you're in the air."

"Right." She looked him over again. There was a lot of dried blood, and that knee was going to need icing, but it didn't look serious. Still, those grazes on his arm and, she noticed for the first time, the palm of his right hand, would burn when she cleaned and disinfected them.

"Isn't that the same trick you were trying last time when you came home looking like this? That can't have been more than a week or so ago."

"Think so." Sean grinned again.

"Wait here and I'll go and get a wet cloth to clean you up a bit. Then we'll take you in, get those dirty clothes off you, and put you in the shower."

"Why do you do it?" she asked as she returned a moment later with a damp hand towel and started to rub him down. It was a statement more than a question, and Sean knew it, but he gave her an answer anyway, familiar with the ritual.

"It's fun," Sean said.

"It can't be fun to keep falling down and hurting yourself," Sally answered.

"No, that's not fun. But it's really cool when you pull it off," he said.

Sally asked, "And have you done a kick flip yet?"

"Not yet. But I will. I just need to keep trying."

She knew that he would.

"When we get you all cleaned up and you've had your dinner, you need to sit down and do your homework. What is it tonight?"

"Math."

Oh, no, Sally thought. *This is going to be a real struggle. I wish he would put the same effort into his math that he does into skateboarding. When he's out with his friends he keeps trying, over and over again, putting up with all the falls and the cuts and the grazes and the bruises, until he gets it right. But the first time he can't see how to solve a math problem, he gives up, convinced he'll never get it. Why is that?*

School, Cool, and Passion

Why indeed? Why do kids—and adults, for that matter—practice one thing over and over again, putting up with any amount of pain and frustration, and yet not show the same determination for something else? And, of particular relevance to readers of this volume, why does mathematics so often not engender the same kind of passion as playing a musical instrument, mastering a particularly challenging video game, or skateboarding?

Sure, mathematics is hard and requires a lot of dedicated practice to master. But the same is true of all those other activities. Some of them, such as skateboarding, involve real physical pain and sometimes blood, which even the most math-averse among us would have to admit are not common consequences of trying to solve a math problem.

The answer, of course, is the one that Sean gave. He kept trying to master the kick flip because it was a cool thing to do. He'd seen videos online of boarders doing it. A couple of his friends could do it. He had looked at the photographs in *Thrasher* magazine and studied the articles that explained how to do it. He wanted to do it because people he regarded as cool did it. It wasn't that he wanted to *be like* some of the people whose skateboarding feats he wanted to emulate; some of them looked like real jerks, with no other interests in life. For him, he knew that skateboarding was just a fun pastime, not a way of life. Both his parents were excellent role models for the benefits of doing well at school and going to college. Still, he did want to master that kick flip.

Sean actually enjoyed school—except, that is, for mathematics. He just couldn't seem to get it. He found the classes boring, and he couldn't see the point.

He was good with computers and could do any arithmetic problem he needed in another class by using a calculator. But most of the material that he encountered in math class seemed so irrelevant. And frankly, the repeated claims of his parents and teachers that what he learned in the math class would be useful to him later did not convince him. He'd heard that for many years now, and so far that "later" day had not arrived.

Many of the successful people he read about or saw on TV seemed to have gotten along just fine without math. Nothing in math class seemed to connect with the world he lived in or the life he might lead as an adult. As far as Sean could see, math was purely a school subject, not really relevant to life in his world.

Besides, math was hard. But then skateboarding was hard, so were some of the video games he played, and so was learning to play the guitar—and Sean was not afraid of taking on a hard challenge.

Sean was a smart kid with lots of interests. He had drive and ambition. With enough motivation, he would probably put the same effort, or more, into mathematics as he did into skateboarding. What would it take to motivate him? The most likely candidates were relevance to his life and his future, a passionate role model, and a cool image. But without sufficient motivation, he would put all his efforts into other subjects, thank you very much.

The Cool Factor

Of course, Sean and his mother are characters I made up. The skateboarding episode is likewise a fiction. But to me, my story says everything there is to say about the importance of a positive and productive disposition for kids to successfully learn mathematics (or anything else for that matter).

So how can we turn things around? No matter how well a mathematics teacher has mastered classroom technique, no matter how good the lesson plans, and regardless of whether the teacher is likable,[1] I think that today's students need more. In particular, they need to know why they are learning mathematics, and it needs an element of cool.

All right, I'll concede that for most students the cool factor won't begin to approach that of skateboarding or guitar playing or social network gaming, but I'd like to at least get math out of the uncool category. The teacher's enthusiasm can go a long way toward achieving that goal. If you've got it (enthusiasm), flaunt it. If you haven't, then you are probably in the wrong profession (but wouldn't be reading this book anyway).

1. I had at least one teacher who, although an excellent instructor, was decidedly unlikable. But I found him to be incredibly inspiring and motivating. He loved mathematics and gave his students as much of his time as they needed. That's why I did not put scare quotes around *excellent instructor* in this footnote.

In addition to enthusiasm, I believe that we can gain much from spicing up lessons with talk about the history of the mathematics being taught, the people who discovered it, why they did so, and how people use it in today's world. Since the training that teachers themselves receive rarely covers those aspects of mathematics, that means the oft-beleaguered teacher must acquire this knowledge for her- or himself. The good news is, this is not only easy to do but also enjoyable. It does not involve learning to do new mathematics, something that is always a challenge.[2] It is simply *reading the story* that was and is mathematics. And all humans enjoy a good story. Plenty of material exists. Searching for "mathematics history" on an online bookstore site returns more than 23,000 titles. Admittedly, many of these results are old or irrelevant, but among the more prominent listings you'll find many great reads full of useful material. When you do this, don't forget to check out the entire range of books written by the authors you come across. Often they have written other books that are just as useful but that don't fall under "mathematics history."

The story of the current uses of mathematics is made more compelling when it includes applications that are *very* current. Doing so is more of a challenge. By their nature, books can describe only things in the past. Unfortunately, the news media, which do focus on today, hardly ever cover stories about mathematics or mathematicians. Still, the Web sites of the American Mathematical Society (www.ams.org), the Mathematical Association of America (www.maa.org), and the Society for Industrial and Applied Mathematics (www.siam.org) have sections devoted to mathematics in the news; the *Wall Street Journal* has the excellent "Numbers Guy" column (blogs.wsj.com/numbersguy); and you can access and download all the audio recordings of my own "Math Guy" contributions to National Public Radio at www.stanford.edu/~kdevlin/MathGuy.html.

And for fostering cool, don't forget the entertainment media. Recent years have seen a whole series of successful movies and plays (e.g., *A Beautiful Mind, Good Will Hunting, The Bank*) and one television crime series, *NUMB3RS*, that involves mathematics and mathematicians.[3]

I keep coming back to the cool factor for a good reason. The popular image of mathematics is decidedly not cool—among any age group of the population.

---

2. We now have some understanding of why mathematics is hard. A huge factor is its highly abstract nature. The human brain did not evolve to handle abstractions. See my book *The Math Gene* (Basic Books, 2000).

3. The Web has several excellent resources for mathematics, some of which refer to current applications; all can lead you to clips that you can profitably show in class. Chapter 16 of this yearbook may be helpful. Also, try the "Math and the Movies Resource List" (mathbits.com/MathBits/MathMovies/ResourceList.htm), "Mathematics in Movies" (www.math.harvard.edu/~knill/mathmovies), and "The Math in the Movies Page" (world.std.com/~reinhold/mathmovies.html).

That means that the mathematics teacher has to not only cover the curriculum proficiently but also supply almost all the motivation to learn the subject. That's a heavy burden to bear. TV and movie clips and fascinating anecdotes can get the teacher only so far.

What else can the math teacher do to generate motivation? And, turning the searchlight on myself, how can those of us in higher education help prepare future math teachers to carry out this motivational role?

The Infection Factor

The answer for both groups is remarkably easy to state, though in my experience it is rarely mentioned. It's not what we know; rather, it's our own attitude to math and how others perceive us. Enthusiasm and positive disposition are highly infectious. They pass from one human being to another whenever the recipient is exposed over time to the one already infected. Neither has to make any particular effort.

We've all experienced this phenomenon. Many have researched and written about it. Try, for example, the books by Robert A. Sullo, such as *The Inspiring Teacher: New Beginnings for the 21st Century* (National Education Association, 1999); *The Motivated Student: Unlocking the Enthusiasm for Learning* (Association for Supervision and Curriculum Development, 2009); or the recent article by David C. Geary, "An Evolutionarily Informed Education Science" (*Educational Psychologist* 43 [October 2008]: 179–95).

Of course, the first person has to already be infected with the disease. Simply trying to fake the symptoms won't work. The recipient may end up being infected, but the virus itself will have to come from another source: the secret to passing on a love and passion for mathematics to students is for the teacher to have that love and passion.

And the secret for producing teachers who have that love and passion is for the colleges and universities that train them to arouse it in them, so that they may then engender it in their students.

Accomplishing this goal doesn't require teaching future teachers to be better mathematicians or to major in mathematics (though either would be good in itself), nor that they receive additional instruction in education theory or educational psychology or have more classroom teaching practice (though those would likewise have value). What I do believe would help is to give future teachers an overview of mathematics as a part of human culture: its nature, its history, and the broad range of applications in today's world. This is the one thing I've mentioned that is most conspicuously absent in the training that our future teachers receive today.

True, most of those applications involve advanced mathematics well beyond the K–12 level. But just as you don't have to play a musical instrument to enjoy listening to music, or paint pictures to enjoy a visit to the art gallery, or play football to appreciate the intricate plays in the Sunday NFL game, so too you can appreciate and enjoy mathematics that you cannot do yourself (or even fully understand).

Supplying this missing ingredient is a key to what I call the cool factor in education. Why "cool"? After all, it's still math that I'm talking about, right? Surely, nothing the teacher does can make math seem cool, at least not for most students.

I disagree. No matter what your age, nothing seems cooler than people with a deep passion for something, who can talk about the object of their passion and demonstrate how it brings them alive and gives them enjoyment, and how it relates to other things in life.

Musicians, painters, actors, athletes, and skateboarders wear their passion on their sleeves for all to see. As a result, they serve as attractive role models. People see them as cool. Mathematicians and math teachers should do the same.

Rocket Science?

Recognizing and implementing the cool factor is not rocket science. We just need to do it. (Actually, rocket science is not that hard, and it is one of the cool, real-world applications of mathematics that a student can master early in a college career. It can easily be explained to future teachers in an understandable way that they can pass on to their pupils.)

Embracing the cool factor does, though, require an adjustment of attitude, by teachers, administrators, and society at large, about what teaching mathematics involves. Let me show you.

I've been giving occasional presentations about these ideas to teacher groups for many years, using a variety of current, real-world applications of mathematics as examples—applications from all walks of life. I always give teachers specific instances of things I think they should show to their students.

Among the cool applications of mathematics I've thus presented are to find answers to these burning questions of the day[4]:

- Why do golf balls have dimples? (It makes them fly up to 2.5 times farther.)
- If the dimples on golf balls make the ball fly farther, why don't airplanes have dimples? (Some do have the mathematical equivalent of dimples.)

---

4. Since these are among my favorites, I have used them myself in various articles and books: *The Math Instinct* (Thunders Mouth Press, 2005), *Life by the Numbers* (John Wiley, 1998), *The Language of Mathematics: Making the Invisible Visible* (W. H. Freeman, 1998), and *The Numbers Behind NUMB3RS: Solving Crimes with Mathematics* (cowritten with Gary Lorden; Penguin-Plume, 2007).

- What really keeps an airplane in the sky? (The answer in most math books,[5] and many pilot instruction manuals, is wrong.)
- Why do honeycombs have a regular hexagon pattern, like a tiled bathroom floor? (The answer is efficiency, but it took mathematicians 2000 years to figure it out correctly. The bees were faster.)
- How many different kinds of soap bubbles can there be? (The mathematics shows that some are theoretically possible but are too unstable to survive long enough for us to ever see them, except on a computer screen.)
- How do birds and fish find their way when they migrate between the seasons? (They are naturally evolved trigonometers.)
- Where does a skateboard get the vertical upward force to leave the ground when its rider executes a jump? (Yes, I sometimes throw skateboarding into the mix. It's much more complicated, and mathematical, than Sean could ever imagine.)
- Why does the supposedly random Shuffle function on the iPod often give you some songs much more frequently than others? (Because it is random.)
- How many different ways can you repeat a design to create wallpaper? (Seventeen.)
- Is it true that the ancient Greeks used the golden ratio to design their buildings, including the Parthenon? (No, it's a myth, along with many other oft-repeated claims about the golden ratio in art, music, and even human physiology.)[6]
- Why do the famous Fibonacci numbers keep turning up in the natural world? (It's another example of nature being efficient, and the mathematics is really neat. This is one case that really involves the golden ratio.)[7]
- Does a baseball outfielder run to catch a ball the same way a dog does? (They both use the peculiarities of the visual system to take advantage

---

5. Including my own *The Language of Mathematics,* cited in the previous footnote.

6. This bizarre belief, for which no one has ever produced a shred of evidence, and which historians have traced back only to the nineteenth century, lives on, though it should have been put to rest by the evidence in the excellent scholarly article "Misconceptions about the Golden Ratio" by George Markowsky, published in *The College Mathematics Journal,* Vol. 23 (January 1992), pp. 2–19, and elaborated on at even greater length by Mario Livio in his book *The Golden Ratio: The Story of PHI, the World's Most Astonishing Number,* Broadway Books, 2003.

7. See the Livio book cited in the previous footnote.

of the same pretty complex math.)

- Where did cycling legend Lance Armstrong find up to eight minutes by which he won each of his seven Tour de France victories? (It wasn't drugs, unless you think that math and science are drugs.)[8]
- Speaking of Lance, just how does a bicycle turn? (The mathematics of this one is much more complicated than you ever imagined. You don't turn a bicycle by turning the handlebar the way you turn a car by using the steering wheel. If you try that, you'll fall down. You do one of two things; either you keep the handlebars straight and simply lean to the side you want to turn, or—and this is the one most of us do, albeit without realizing it—you turn the handlebars the opposite way from the direction you want to turn, a maneuver called countersteering. This concept is counterintuitive. You need to think mathematically to understand why things work this way.)
- Oh yes, rocket science.

After I give these talks, the organizer usually gives me a copy of the anonymous feedback forms the audience has filled in. The most common response I get goes something like this: "This is interesting stuff, and I enjoyed the talk and appreciated the presenter's obvious enthusiasm, but he didn't go into the details, and besides, the mathematics is way beyond the ability of my pupils."

Indeed it is. A lot of what I present is beyond *my* ability—or at least is mathematics that I do not know how to do myself. But being able to *do it* is not what is required, of either the teacher or the pupil.

What those teachers' comments reflect is that we have allowed ourselves to be conditioned into thinking that learning mathematics is all about learning how to "do math"—to solve math problems. Yes, that is part of learning mathematics, and it's an important part—arguably the most important part. But it should not be the only part. Wanting to learn mathematics greatly facilitates learning how to do it, and that requires the other things I've talked about. They should be part of the educational package as well.

Mathematics teachers should spend a significant part (though not most) of their time teaching students *about* the subject in much the same way that teachers do in history, social studies, or literature. To facilitate this, the K–12 curriculum needs to change to allow for, and require, this crucial additional component,[9] and our colleges and universities need to prepare future and current teachers to teach it.

---

8. The Discovery Channel has an excellent documentary called *The Science of Lance Armstrong*.

9. For effect, I'm focusing here on easily recognizable stereotypes. I know that many teachers would like to include such "extracurricular" materials in their classes, and many

As I say, it ain't rocket science. It's common sense. Remember Sean? Just three days after his fall, as soon as the swelling in his knee had subsided, he was back on his skateboard practicing his kick flip. Nothing was going to stop him from mastering that cool maneuver. He had the motivation (he could see how to apply his skill to do something he and his friends saw value in), he was learning by doing (or trying to do) the thing he was trying to master, and every attempt was followed by immediate, and often painful, feedback. Sure, learning and using mathematics is a lot more complicated than skateboarding, in many ways. Nevertheless, I think Sean has a lesson for us.

Some Additional Sources of Spice

The following magazines (which publish both print and online versions) offer good topical ideas: *The Mathematical Intelligencer, Math Horizons, Science, Science News, New Scientist, Discover, Chance,* and *Significance*. Several organizational online sources, are useful, including *MAA Online, Visual Mathematics, Journal of Mathematics and the Arts,* and *Journal of Mathematics and Music.* You can also find biographies of leading mathematicians throughout the ages to the present day at the University of St. Andrews's "Indexes of Biographies" (www-groups.dcs.st-and.ac.uk/~history/BiogIndex.html). For books, start with the "popular accounts" of mathematics by Ian Stewart, William Dunham, Amir Aczel, Mario Livio, Ivars Peterson, John Allan Paulos, Clifford Pickover, and me. Enjoy!

(Keith Devlin is a mathematician at Stanford University, where he is a cofounder and director of the Human-Sciences and Technologies Advanced Research Institute [H-STAR]. His research interests involve using different media to teach and communicate mathematics to various audiences and using mathematical ideas in intelligence analysis. He has published around eighty research articles and twenty-eight books, many for a general audience. He is "the Math Guy" on National Public Radio.)

do. I often meet them, and sometimes they e-mail me for suggestions of suitable "enrichment materials." But finding a spare moment in the crowded curriculum is not the same as having it be a recognized and important component of mathematics education.

Chapter 18

Venturing into Storybooks: Motivating Teachers to Promote Mathematical Problem Solving

Jane M. Wilburne
Jane B. Keat

PROMOTING MATHEMATICS instruction to prepare young learners to develop as mathematical thinkers and to become confident in their mathematical abilities is important. The National Association for the Education of Young Children (NAEYC) and the National Council of Teachers of Mathematics (NCTM) published a joint position statement that highlights the need for primary-grade teachers to emphasize effective mathematics instruction for young children through teaching practices that strengthen children's problem-solving and mathematical reasoning abilities (NAEYC and NCTM 2002). The joint statement also recommends that teachers engage young children in deep and sustained interactions with key mathematical ideas and teach mathematics by integrating it with other activities. Although many publications have supported integrating literature with the teaching of mathematics, many references focus on using such books as a "springboard" or "tool" to introduce children to the mathematics (Kliman 1993; Welchman-Tischler 1992; Whitin and Wilde 1992; Schiro 1997). After presenting the mathematics, teachers no longer use the storybook in the instruction. This article will focus on a unique perspective for integrating mathematics into storybooks and the pedagogical transformations that primary-grade teachers experienced in doing so. Specifically, this paper gives a framework showing the phases of motivation that teachers who used storybooks to promote mathematical problem solving in their classrooms experienced and summarizes the process for use in any primary-grade classroom. This process or an adaptation of it may serve teachers of older students as well.

Using Storybooks to Find the Mathematics

Teaching mathematics through storybooks has helped young children make connections between mathematics in the classroom and lifelike situations and motivated them to explore mathematical problems (Hong 1996; Ducolon 2000;

Wilburne et al. 2007). Although an abundance of storybooks is available to teachers, the technique of using such books to teach mathematics is often ignored or remains at the periphery of the mathematics curriculum (Hunsader 2004). Many primary-grade teachers eloquently express joy when they read storybooks to the children in their classes, yet they do not think of using the same storybooks to set the stage for mathematical exploration and problem solving (Wilburne et al. 2007). Even though many resources that list storybooks and the mathematical concepts they address are available to teachers, teachers need guidance in how to use the storybooks to launch mathematical conversations and problem-solving explorations (Wilburne et al. 2007).

In our work with primary-grade teachers, we found that many expressed a lack of confidence in their ability to find mathematical problems within a storybook and pose open-ended problems without access to a textbook teachers' guide. The teachers stated that they typically used concept books that emphasized counting, numbers, or specific mathematical topics as one way to link literacy with mathematics. Concept books typically are informational books that include pictures, labels, and captions. Books that teachers used included *12 Ways to Get to 11* (Merriam 1993), *The Coin Counting Book* (Williams 2001), and *The Jelly Bean Fun Book* (Capucilli 2001). The teachers noted that when they used concept books, they found minimal opportunities to pose problems that were open ended or required students to use different strategies. They had not thought to use storybooks that have a plot, character(s), and a setting to intellectually involve students in thinking about mathematics, physically involve them in doing mathematics, and emotionally involve them with the impact that the book's mathematics might have on its characters or on the child's own life (Schiro 1997). If the mathematics wasn't obvious in the story, the teachers failed to see how to integrate the book into a mathematics lesson.

The teachers were surprised when we asked them to find mathematical problem-solving opportunities in storybooks such as *Minnie's Diner: A Multiplying Menu* (Dodds 2007) or *Who Sank the Boat* (Allen 1982). They stated some obvious concepts such as counting the plates of food or counting the animals in the boat. But when we asked them to pretend that they were in the story—that they were a character in the story, and they had to imagine some mathematical problems that the character(s) might pose—they suddenly looked stunned. They had never thought to see the mathematics from a character's point of view or from within the story. They started thinking creatively and found that they could pose mathematical problems that used characters from the story, the storybook illustrations, or the story plots to frame the context of the problems (figs. 18.1 and 18.2).

Mathematical Problems Posed before Learning the Venturing Technique

- What does double mean?
- What do you get when you add 4 + 4, 8 + 8, and 16 + 16?
- Which son ate the most food?
- What pattern do the numbers make?
- How many food items does the waitress bring out to the son named Bill?

Mathematical Problems Posed after Using the Venturing Technique

Papa sees his very large shadow.

- How could my shadow be so large?
- How long is my shadow?
- Can your shadow be bigger than you are? How could we find out?
- Bill wants to know how long it will take each boy to eat his meal. How could he find out?
- The cook needs to know how many loaves of bread he will need to make all the sandwiches in the story. How could we help him figure this out? Let's draw pictures of sandwiches to help us.
- Gill is chopping wood. He can chop down four trees in an hour. How long would it take him to chop down twelve trees? Let's use the colored chips to figure this out.
- If Gill is 6 feet tall, how much taller than you is he? How could we find out? If the door frame is 62 inches high, can he fit through the door?
- The waitress is wondering how heavy the tray of food on the last page is. Do you think the tray weighs more or less than five pounds? How could she find out?

Fig. 18.1. Sample of a second-grade teacher's mathematical problems created from *Minnie's Diner: A Multiplying Menu* (Dodds 2007)

The teachers began to see the potential of linking mathematics to children's ability to pretend and imagine along with the storybook's characters and context. They became motivated to find the math in the stories and develop problems involving mathematical concepts connected to everyday life such as time, measurement, geometry, data analysis, numbers, and probability. Eventually, they began selecting storybooks and enthusiastically looking for mathematical concepts that they could address through the story and problems they could pose to engage their students in problem solving.

Mathematical Problems Posed before Learning the Venturing Technique

- How many animals are in the boat at the end?
- Which animal do you think weighs the most? The least?
- What does *balance* mean?
- What do you notice every time another animal gets into the boat?

Mathematical Problems Posed after Using the Venturing Technique

- If the mouse decided to get into the boat first, what do you think would happen? When would the boat sink?
- If the bay was 6 feet deep, could the horse touch the bottom of the bay? Why or why not? Could you touch the bottom of the bay?
- How large do you think the bay in the story is? How could someone measure the size of the bay?
- When all the animals are in the boat, how many feet are in the boat? How many ears are in the boat?

Fig. 18.2. Sample of a first-grade teacher's mathematical problems created from *Who Sank the Boat* (Allen 1982; cover reproduced with permission of Penguin Group USA)

As Vygotsky (1978) stated, children who are engaged in conversations within make-believe situations can think with forms of complex thought that they cannot yet enact in realistic situations. He suggested that children can do in play what they cannot do in real life. We found that teachers became more motivated to find the mathematics in storybooks when they realized how excited their students became when solving mathematics problems based on imaginary contexts from stories. Further, the teachers discovered how storybooks with playful and imaginative language could lend themselves to authentic and purposeful mathematical experiences for their students. They also realized how using storybooks to frame mathematical discussions helped their students see that mathematics is not just a topic to think about during math time or only when using a math workbook.

Motivation with Storybooks

One definition of motivation is that it explains "*why* people decide to do something, *how hard* they are going to pursue it, and *how long* they are willing to sustain the activity" (Dörnyei and Ushioda 2010, p. 4). Using storybooks in teaching mathematics motivated the teachers to find the mathematics in various storybooks. When teachers taught mathematics through the characters or the context of the stories, they were more motivated to repeatedly use the same storybooks to emphasize different mathematical concepts or strategies. They stated that reading through the story first was better because the students were intrigued and wanted to see what happened. Then they repeatedly revisited the story and read a short section of the book, stopped, and posed an "I wonder" or a "what if" mathematics question. These questions would then lead into the mathematical discussion of the day. For example, one teacher noted that the storybook *A Chair for My Mother* (Williams 1982; cover reproduced with permission of HarperCollins Publishers) stimulated class conversations about comparative weights of a jar with different quantities of coins that family members collected to replace a chair that was lost in a house fire. In another reading, the teacher asked the class to wonder about the cost of furniture items such as those in the storybook. She had the students estimate the costs and then use store circulars from the newspaper to find furniture prices to practice reading and writing large numbers.

As the teachers used various storybooks, they became more skilled in posing open-ended problems. They posed more problems that required their students to make connections across mathematical topics, and the teachers began to see how easily they could excite their students to want to explore mathematical relationships and concepts. Once they had success with one storybook, the teachers were more motivated to select another book and find the mathematics in it.

Emphasizing the Curriculum Focal Points

Venturing into storybooks helped teachers frame problems around the content emphases as stated in the *Curriculum Focal Points for Prekindergarten through Grade 8 Mathematics: A Quest for Coherence* (NCTM 2006). Teachers found having a list of the focal points to use as a reference helpful when deciding how to use the storybooks as a context for the mathematics lesson. In one kindergarten classroom, the teacher identified the focal point defined as ordering objects by a measurable attribute. She selected the storybook *The Napping House* (Wood 1984), which is a cumulative tale about a snoring granny who is sleeping on a rainy afternoon. Gradually, various critters crawl on top of Granny (e.g., a child, a dozing dog, a snoozing cat) as she sleeps. The story ends when a flea sets off a chain of events that results in a pile of characters and a broken bed. The teacher began with questions that dealt with counting the critters or number of legs on the bed each time. Then she focused questions around the focal point, for example, "Which animal weighs more?" "Which animal weighs less?" "How do you know?" "How could we find out?" "What size bed would the granny need to fit all the critters?" She asked students to pretend that each critter brought a friend on the bed, and she asked, "How many critters would be on the bed now?" "Let's pretend the bed is one foot long; can you show me with your hands how long the bed would be?" (This was followed up by comparing hand measures with a ruler to demonstrate the length of a foot.) "Can you find something around the classroom that would be longer than the bed? Let's measure it." The discussions then continued with students measuring various objects in the classroom by using nonstandard units and comparing results.

For second grade, *Curriculum Focal Points* addresses three content emphases: developing an understanding of the base-ten numeration system and place-value concepts, developing quick recall of addition facts and related subtraction facts and fluency with multidigit addition and subtraction, and developing an understanding of linear measurement and facility in measuring lengths. One teacher used play food to model the addition of items on the tray that the waitress in *Minnie's Diner* (Dodds 2007) carried. She had the students pretend that they were at the restaurant and that they were going to order some of the food Minnie carried. They had a list of the prices of the food on a chart at the front of the classroom, and students used their base-ten blocks to solve multidigit addition problems to determine how much their meals would cost them. Then they recorded their problems and drew pictures to represent the food that they were ordering. In another classroom, a teacher had students using square tiles to determine how many customers could sit along the perimeter of the table if each side could fit one person. Next, he gave them two, three, and four squares and asked them to explore the number of customers who could be seated if the tables were placed side by side. Finally, he had them create a chart with this information to look for patterns between the number of square tables and the number of customers that could be seated. This activity branched into algebraic thinking—with no mention of algebra in the story. Making connections between the grade-level Focal Point and the storybook motivated these teachers to create their own open-ended mathematics problems.

Framework for Phases of Motivation

As we worked with various primary-grade teachers to use storybooks to promote mathematical problem solving, we saw a recurring series of phases that the teachers experienced as they became more and more motivated to teach mathematics differently from their traditional approach. Hence, we offer a framework showing the transformation of teaching mathematics with storybooks. The framework describes three phases (fig. 18.3) that teachers experienced in their efforts to use storybooks to engage students in mathematical problem solving. As teachers moved through each phase, they became more motivated to select storybooks and find the math.

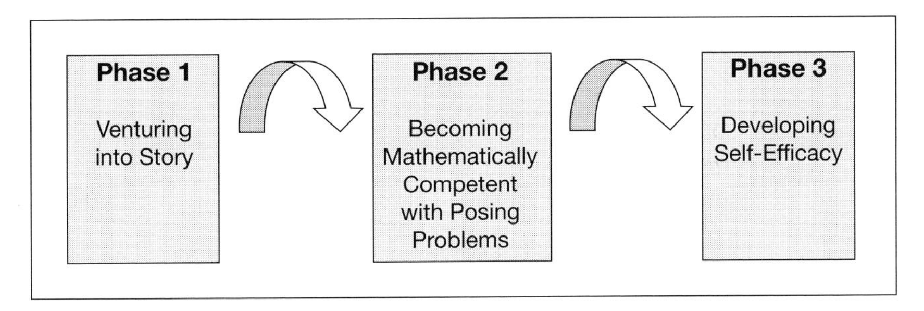

Fig. 18.3. Phases of teacher motivation

Phase One: Venturing into Story

In this first phase, teachers began by selecting a children's storybook that had characters and a plot that could relate to the concept(s) of the mathematics lesson or ones that interest students. They were invited to pretend to either enter into the world of the storybook or take the characters out of the story and place them in the real world. For example, *Benny's Pennies* (Brisson 1995) is about a boy deciding what to buy with his five pennies. He receives suggestions from his mother, his sister, his dog, and his cat, and he ends up buying something for each of them. At first, we asked the teachers to pretend that Benny appeared in their classroom and to think about what mathematical questions he might ask.

Throughout this phase, teachers began to consider various contexts and mathematical problems that they could face in the storybook world or problems that the storybook characters could encounter. By relying on young children's innate capacity to pretend, teachers found that they could take story characters out of the books and into the children's bedrooms, kitchens, or community to create mathematical problems. This way of thinking helped the teachers pose more "what if" questions to launch students' mathematical thinking (e.g., "What if Bennie had a dime?" "What if Bennie saved one penny every day for a week?").

The teachers began to create problems that did not always have a single answer and required students to engage in higher-order thinking. Many primary-grade teachers stated that they rarely created their own mathematics problems because they were afraid that they themselves may not know the answer. By venturing into the story, they found that posing open-ended problems for their students was easier than before and noted that they "found a sense of freedom that allowed them to think outside the textbook." Now they enjoyed the challenge of taking any storybook and picturing one of the characters visiting the classroom or pretending that they lived in a page from the story as a context for a mathematical problem.

Phase Two: Becoming Mathematically Competent with Posing Problems

In phase two, the teachers began to feel more competent in creating problems that were related to the mathematics curriculum and that could extend the mathematical thinking of their students. The teachers could pose mathematical problems that were more authentic and relevant to the children's everyday lives. They created problems around concepts of time, measurement, geometry, data, number, and probability: for example, "Suppose that Benny wanted to play a game called 'flipping the penny.' Every time the penny landed on heads, you receive one penny, and every time it landed on tails, Benny receives two pennies. Would this be a fair game?"

Many problems that they created required students to apply various problem-solving strategies such as guess and check, making a list or table, drawing a diagram, acting it out, or working backwards. Teachers began to see how problem solving is more than solving the typical word problems found in some mathematics textbooks. They used storybook illustrations, storybook characters, and storybook plots to create contexts for problems that aligned with state standards and the NCTM *Curriculum Focal Points*.

As teachers expressed more comfort with venturing into the story, they became more competent in developing open-ended problems that were mathematically meaningful to their students and engaged their students in mathematical explorations and discoveries. They began to see mathematics as real and non-threatening and how it is more than just having students perform operations or memorize facts.

Phase Three: Developing Self-Efficacy

In phase three, teachers began to feel confident in veering away from relying on a textbook with an answer key, and they recognized that they could stimulate their students' mathematical thinking through problem solving. They realized that mathematical problems exist everywhere, and they started to pose more open-ended problems both with and without storybooks. The teachers began to develop a sense of self-efficacy and the belief that they could think mathematically and pose questions even if they did not know the answer.

Ashton (1985) defines teachers' sense of efficacy as "their belief in their ability to have a positive effect on student learning" (p. 142). Before venturing into the storybooks, many teachers made comments such as "I only do the mathematics in the textbook" and "I was never very good with problem solving, so I know I don't do enough with my students."

After having some experience creating problem-solving situations from the storybooks, many teachers began to have a sense that they could find the math. "I can do this!" exclaimed one first-grade teacher after seeing the process modeled and creating problems from a storybook. Another first-grade teacher stated, "I see more math around me now and pose more math questions when reading storybooks." She described how her students seemed to scaffold her awareness of mathematics beyond the storybook *Minnie's Diner*. As she asked students what they could carry on a tray, Tommy imagined that he could "carry a tray with two items, a large cup of water and a sandwich" (fig. 18.4). Simultaneously, Sharisa pointed out that she loved ham sandwiches and would buy fifty if she were in

Fig. 18.4. Tommy's work on *Minnie's Diner*

Minnie's Diner (fig. 18.5). Connor made a connection to the storybook by recalling the details of what happened to him in a restaurant: he lost a tooth (fig. 18.6). The teacher saw opportunities to extend students' mathematical knowledge by inviting them to explore ideas of fitting items on a tray, consider concepts of relative costs, and think about adding and subtracting teeth.

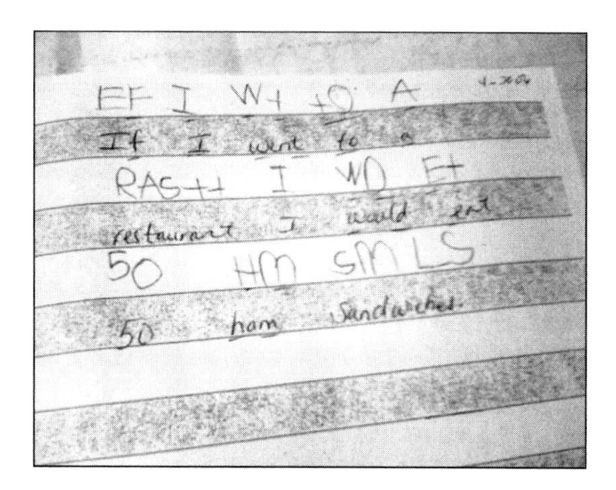

Fig. 18.5. Sharisa's work on *Minnie's Diner*

Fig. 18.6. Connor's work on *Minnie's Diner*

Impact on Students' Mathematical Learning

Using children's storybooks positively affects students' ability to communicate mathematics and to use language to help them and others construct mathematical meaning and develop common mathematical understandings (Griffiths and

Clyne 1991; Hinton 1991). Our work found that when teachers created imaginary situations and circumstances that storybook characters encountered, their students were more excited to solve the problems posed and used more mathematical language in their kid writing. The students had opportunities to think like mathematicians and used reasoning and higher-order thinking skills. In several kindergarten classrooms, using storybooks to teach a unit on money led to significant test score gains (Wilburne et al. 2007). This finding supports the work of Hong (1996) as well as that of Jennings and colleagues (1992), who showed that children's mathematical achievement improved when teachers taught mathematics with storybooks.

Having a positive experience with the mathematics that develops from an interesting story also motivated the students to reread the stories, act out mathematics problems, and make mathematical connections to their world and the world of the storybook. Several kindergarten teachers noted how their students were more motivated to play "math" and act out mathematical problems with the storybooks than when the mathematics instruction was based solely on a textbook. For example, some children talked about coins, played with the coins in the classroom center's bank, and drew marks to represent how many cents each coin represented. They were doing so not as a random exercise but as a character from the story *Benny's Pennies* (Brisson 1995) who needed to know about coin values to buy items for his family. In another classroom, students were rereading *A Chair for My Mother* (Williams 1982) and counting base-ten blocks by tens to represent how many dimes they would need to buy certain items for their family from a catalog. The teachers always found that the storybooks and storybook characters spoke to the heart of the child and offered a much more personal way to learn mathematics.

Conclusion

This article highlights how storybooks motivate primary-grade teachers to make mathematical connections, promote more reasoning and problem solving, and enhance the level of mathematical communication in their classrooms. Through using storybooks, teachers find the ease with which they can accomplish this goal and how they can infuse the pedagogical approaches that *Principles and Standards for School Mathematics* (NCTM 2000) emphasizes and the *Curriculum Focal Points* (NCTM 2006) content emphases. Figure 18.7 summarizes the process that any classroom teacher can use to venture into storybooks.

1. Select storybooks with characters and a plot that could relate to the concept(s) of the mathematics lesson or ones that interest students.
2. Read the storybook cover to cover to understand the story and the characters.
3. Identify the state standards or Curriculum Focal Points (NCTM 2006) relevant to the curriculum that can supply a context for mathematical problems.
4. Select a scene from the storybook and imagine yourself entering into the scene. What do you notice? What mathematical problems could you ask yourself in that scene?
5. Select a character from the story and imagine the character coming to life and entering your classroom (or kitchen, etc.). What mathematical questions could this character ask? What might this character be wondering? Think about different open-ended questions that the character could pose, whether there is a known answer.
6. Develop problem-posing questions that align with the mathematical concepts that you are teaching. Write problems on sticky notes and insert them into the storybook. Use various problems at different readings of the storybook to engage students and build their interest in the story and the mathematics.
7. Use the different readings of the storybook to focus on specific mathematical content areas such as number and operations, geometry, measurement, probability, data analysis, and algebra.
8. Create journal prompts based on a mathematical problem that one of the storybook characters posed, to which students respond in writing or orally.

Fig. 18.7. Venturing into storybooks

Teachers began within the comfort zone of using concept books or their traditional mathematics textbooks, and they gradually progressed to taking the characters from storybooks into the land of imagination to explore ways to promote more problem solving in their classrooms. We saw how teachers posed problems that took the students and storybook characters to places such as the moon, to the depths of the ocean, and even to the tops of trees in a Canadian forest. The teachers were more excited to teach mathematics when they created open-ended problems based on their students' tendencies to pretend and imagine. Thus, the storybooks motivated the teachers, who further motivated the students to engage in mathematical thinking enthusiastically.

REFERENCES

Allen, Pamela A. *Who Sank the Boat?* New York: Putnam Berkley Group, 1982.

Ashton, Patricia T. "Motivation and Teachers' Sense of Efficacy." In *Research on Motivation in Education, Vol. 2: The Classroom Milieu,* edited by Carole Ames and Russell Ames, pp. 141–74. Orlando, Fla.: Academic Press, 1985.

Brisson, Pat. *Benny's Pennies.* Illustrated by Bob Barner. New York: Dragonfly, 1995.

Capucilli, Karen. *The Jelly Bean Fun Book.* New York: Little Simon, 2001.

Dodds, Dayle Ann. *Minnie's Diner: A Multiplying Menu.* Illustrated by John Manders. Cambridge, Mass.: Candlewick Press, 2007.

Dörnyei, Zoltán, and Ema Ushioda. *Teaching and Researching: Motivation.* 2nd ed. New York: Pearson Longman, 2010.

Ducolon, Colin K. "Quality Literature as a Springboard to Problem Solving." *Teaching Children Mathematics* 6 (March 2000): 442–46.

Griffiths, Rachel, and Margaret Clyne. *Books You Can Count On: Linking Mathematics and Literature.* Portsmouth, N.H.: Heinemann, 1991.

Hinton, Jacki. *Kindercorner Math: Linking Children's Books to Math.* Littleton, Mass.: Sundance, 1991.

Hong, Haekyung. "Effects of Mathematics Learning through Children's Literature on Math Achievement and Dispositional Outcomes." *Early Childhood Research Quarterly* 11 (December 1996): 477–94.

Hunsader, Patricia D. "Mathematics Trade Books: Establishing Their Value and Assessing Their Quality." *Reading Teacher* 57 (April 2004): 618–29.

Jennings, Clara M., James E. Jennings, Joyce Richey, and Lisbeth D. Krauss. "Increasing Interest and Achievement in Mathematics through Children's Literature." *Early Childhood Research Quarterly* 7 (June 1992): 263–76.

Kliman, Marlene. "Integrating Mathematics and Literature in the Elementary Classroom." *Arithmetic Teacher* 40 (February 1993): 318–21.

Merriam, Eve. *12 Ways to Get to 11.* Illustrated by Bernie Karlin. New York: Simon and Schuster, 1993.

National Association for the Education of Young Children (NAEYC) and National Council of Teachers of Mathematics (NCTM). *Early Childhood Mathematics Education: Promoting Good Beginnings.* Washington, D.C.: NAEYC and NCTM, 2002.

National Council of Teachers of Mathematics (NCTM). *Principles and Standards for School Mathematics.* Reston, Va.: NCTM, 2000.

———. *Curriculum Focal Points for Prekindergarten through Grade 8 Mathematics: A Quest for Coherence.* Reston, Va.: NCTM, 2006.

Schiro, Michael. *Integrating Children's Literature and Mathematics in the Classroom.* New York: Teachers College Press, 1997.

Vygotsky, Lev S. "Interaction between Learning and Development." In *Mind and Society: The Development of Higher Psychological Processes,* edited by Michael Cole, Vera John-Steiner, Sylvia Scribner, and Ellen Souberman, pp. 79–91. Cambridge, Mass.: Harvard University Press, 1978.

Welchman-Tischler, Rosamond. *How to Use Children's Literature to Teach Mathematics*. Reston, Va.: National Council of Teachers of Mathematics, 1992.

Whitin, David, and Sandra Wilde. *Read Any Good Math Lately?* Portsmouth, N.H.: Heinemann, 1992.

Wilburne, Jane M., Mary Napoli, Jane B. Keat, Kim Dile, Michelle Trout, and Susan Decker. "Journeying into Storybooks: A Kindergarten Story." *Teaching Children Mathematics* 14 (November 2007): 232–37.

Williams, Rozanne Lanczak. *The Coin Counting Book*. Watertown, Mass.: Charlesbridge, 2001.

Williams, Vera B. *A Chair for My Mother*. Illustrated by Vera B. Williams. New York: Greenwillow, 1982.

Wood, Audrey. *The Napping House*. Illustrated by Don Wood. San Diego: Harcourt Brace, 1984.

Chapter 19

TARGETTS: A Tool for Teachers to Promote Adaptive Motivation in Any Mathematics Lesson

Melissa C. Gilbert
Lauren E. Musu-Gillette

MANY TEACHERS ARE concerned about how to positively and effectively motivate their students in mathematics while also attending to state (e.g., California State Board of Education 1999) and national (e.g., National Council of Teachers of Mathematics [NCTM] 2000) standards (e.g., Silver et al. 2009). Although theories of motivation are typically part of the educational psychology courses that preservice teachers take, motivation researchers have long argued that teacher candidates rarely receive a comprehensive discussion of how these psychological theories relate to many topics crucial to teachers, such as classroom management, individual differences, and high-stakes testing (e.g., Ames 1990). At the same time, in-service teachers whom we encounter at professional development workshops often ask for effective strategies for motivating their students (e.g., "how to motivate students who have absolutely no interest in learning at all" and "how to motivate my gifted students and what to do with my lowest students who don't get math at all").

In this paper, we share specific classroom strategies that enhance student motivation and dispositions toward mathematics while also promoting student mathematical understanding. We begin by defining "adaptive motivation," including both theoretical and empirical support for three aspects of positive motivation (value, confidence, and mastery achievement goals). Next, we share the eight dimensions of TARGETTS, an analytic framework rooted in motivational research (e.g., Epstein 1988; Ames 1990; Maehr and Anderman 1993) that highlights specific instructional strategies that can support both adaptive motivation and mathematical understanding. Finally, we show how teachers can use TARGETTS as a tool for reflection to evaluate and modify existing lessons to promote adaptive motivation.

Understanding Motivation

When we talk with teachers about motivation, we first ask them for some reasons why their students might be motivated to learn mathematics or to avoid doing mathematics. Some teachers begin by telling us what it looks like to be motivated to learn mathematics (e.g., make eye contact, volunteer to participate, complete assigned tasks) or to avoid doing mathematics (e.g., pass notes, stare out the window, groan, come to class unprepared). We encourage them to think about the underlying reasons for the outwardly observable behavior.

Some of the reasons they tell us that their students might be motivated to learn mathematics include that "it is important for their future aspirations" or because "they want to look smart in front of their peers." Their students might be motivated to avoid doing mathematics because they think "they aren't good at it" or they are concerned about looking dumb in front of their peers, saying, "If I don't even try to solve the problem, then no one will know that I don't understand what I'm supposed to do."

Not surprisingly, teachers' responses are consistent with the literature on academic motivation, particularly in mathematics. Several theories of motivation converge on general reasons or motives for behavior (e.g., expectancy–value theory [Eccles et al. 1983] and achievement goal theory [Ames and Archer 1988]). In general, when people consider acting in a situation, answers to the following three questions influence their behaviors:

1. "Can I do this?" addresses whether students feel confident in their ability to do the mathematics tasks their teachers have given to them.
2. "Do I want to?" addresses whether students find value or meaning in the tasks. Do they feel that their teacher is asking them to do something that is interesting, worthwhile, or useful?
3. "Why do I do it?" is about personal reasons for doing the work. Sometimes students are motivated to do something in order to deepen their level of learning and understanding (*mastery achievement* goals), other times they want to show others what they can do (*performance-approach* goals), and still other times they want to avoid a situation that would make them feel stupid (*performance-avoid* goals).

Defining Adaptive Motivation

Although students might be motivated for many reasons, not all of these are consistent with promoting long-term interest and achievement in mathematics. Findings from many research studies have led to defining *adaptive* motivation as students' (1) having a high *self-confidence* in their ability to do mathematics, thinking, "Yes! I can do this problem" (e.g., Wigfield and Eccles 2000); (2) *valuing* mathematics, seeing mathematics as useful and interesting (e.g., Middleton and Spanias 1999; Eccles and Wigfield 1995); and (3) approaching mathematics class with high *mastery* and low *performance-avoid achievement goals*; that is, they are more focused on learning and understanding the material than on avoiding looking dumb (e.g., Anderman and Wolters 2006; Turner and Patrick 2004).

The literature on performance-approach goals is more mixed. Although interpersonal comparisons and competition may have short-term gains for some students, this type of focus in the classroom appears to have long-term detrimental effects on students' engagement and achievement in mathematics (Midgley, Kaplan, and Middleton 2001). Further, teachers can better promote diverse students' learning by encouraging them to work together as a team to deepen everyone's understanding of mathematics. As with a successful sports team, each student in the class contributes to the outcome and everyone has something to improve on through effort and practice. The outcome of the contest depends on the contributions of everyone involved, and mistakes are viewed as learning opportunities for the whole team. Thus, research suggests that, to support students' adaptive motivation, teachers use strategies focused on minimizing endorsement of performance-approach achievement goals.

Importance of Adaptive Motivation

Why do we want our students to have adaptive motivation? Students with higher adaptive motivation tend to do better in math, both in course and test performance and in their ability to persist in the face of challenging work (e.g., Wigfield and Eccles 2000; Middleton and Spanias 1999). Figure 19.1 shows the relationship between adaptive motivation and standardized test performance (SAT-10) for a sample of 568 eighth-grade African American (31 percent) and white (60 percent) students participating in the TEAM-Math project (discussed in chapter 21 of this Yearbook). This chart shows that high achievers reported significantly higher adaptive motivation than low achievers. High achievers were more self-confident in their mathematics ability, valued the subject, and endorsed mastery more highly than students with performance-avoid achievement goals. Results by ethnicity were similar, though the differences were more pronounced for African American students, and white high and low achievers did not significantly differ on endorsement of mastery goals.

Adaptive motivation also relates to students' ability to demonstrate adaptive reasoning, a type of mathematical knowledge entailing a deeper conceptual understanding because it expects students to reflect on, explain, and justify their mathematical process and solution(s) (Kilpatrick, Swafford, and Findell 2001).

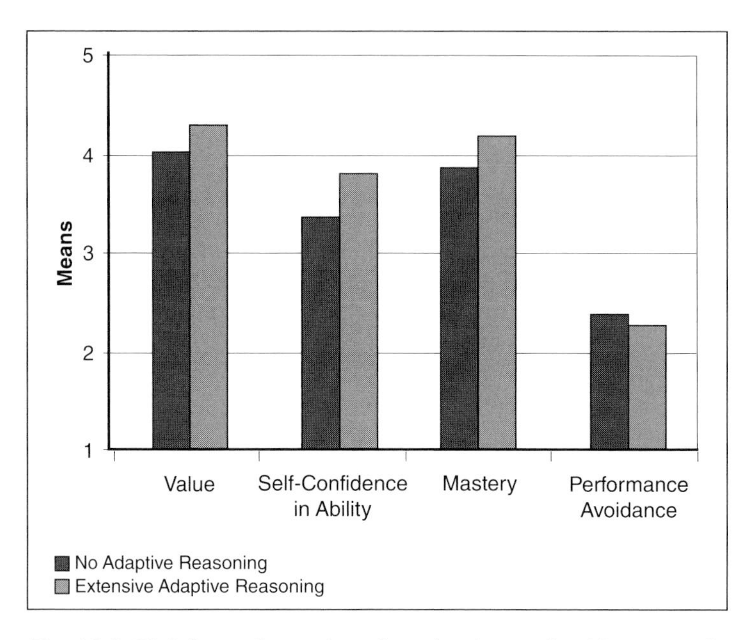

Fig. 19.1. Eighth-grade students' motivation and achievement in mathematics

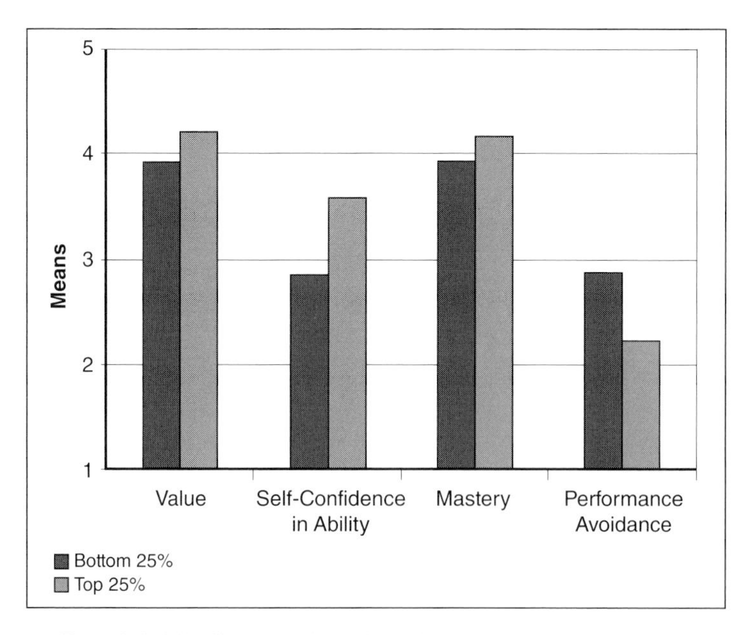

Fig. 19.2. Middle school students' motivation and adaptive reasoning in mathematics

Figure 19.2 summarizes the results for adaptive reasoning from a separate study of 479 prealgebra students (95 percent seventh graders, 69 percent Latino) in Southern California. This chart shows that adaptive motivation differed between students who showed the highest and lowest levels of adaptive reasoning (Gilbert 2008). In sum, students with adaptive motivation are more successful than their peers for many desirable outcomes, including depth of mathematical understanding and standardized test performance.

Dimensions of TARGETTS

Once researchers had defined adaptive motivation, they set out to identify specific instructional practices that supported or hindered students' adaptive motivation. Epstein (1988) identified six dimensions that resonated within the motivation research community (e.g., Ames 1990; Maehr and Anderman 1993) and that have been applied to classrooms and schools: the design of student *tasks*, the level of *autonomy* or responsibility afforded students, the *recognition* of student achievement and learning, teachers' *grouping* of students, the *evaluation* of student work, and the allocation of *time* in the classroom (i.e., the acronym TARGET). Over time, classroom-based research (e.g., Patrick et al. 2003; Stipek et al. 1998; Patrick, Anderman, and Ryan 2002) suggested the need to highlight two additional dimensions: *teacher expectations* for student behavior and performance and teachers' construction and facilitation of *social interaction* in the classroom (i.e., TARGETTS).

Within each TARGETTS dimension are decisions that teachers make that prior research suggests would promote or diminish adaptive motivation. For the task dimension, teachers support students' adaptive motivation when they make problems relevant to students' daily lives and encourage them to show how they got their answer (Middleton 1995). Teachers diminish students' adaptive motivation when they assign many disconnected, repetitious problems and emphasize competition between students, for example, by presenting the task as a race to see who can finish first (Turner and Patrick 2004). For another example, consider recognition, or *why* students are rewarded. Teachers promote students' adaptive motivation when they reward students' thinking processes at least as much as the correctness of the final answer (Turner et al. 1998). Teachers diminish students' adaptive motivation when their focus is on who got the right answer and how quickly (Stipek et al. 1998).

The interrelationships among the TARGETTS dimensions help teachers to ensure that they are sending consistent motivational messages to their students. For instance, decisions about recognition are necessarily intertwined with evaluation, that is, grading and reporting. During class discussions, teachers may validate that students' solution approaches are important, but then, on the chapter test, fail to allot space or award points for students' work. Or teachers might say that they are interested in each student's personal growth in mathematics but then count only end-of-unit assessments in their gradebook, without considering how much students have learned since the pretest. These mixed messages tend to diminish students' adaptive motivation, so we encourage teachers to use TARGETTS as a sort of "reality check" across their instructional practices.

Many strategies that TARGETTS encourages teachers to use are also "good teaching" according to the *Principles and Standards for School Mathematics* (National Council of Teachers of Mathematics [NCTM] 2000). This is not surprising since these practices were specifically recommended because they support students' adaptive motivation and because adaptive motivation is positively related to achievement. Unfortunately, analyses of mathematics classroom practice show that typical instruction is not consistent with the NCTM Standards (e.g., Jacobs et al. 2006). When teachers are under pressure to cover a given amount of content in a short time, they tell us that their inclination is to use more traditional didactic approaches to save time. But rushing through material tends to have the opposite effect, diminishing students' adaptive motivation and limiting their opportunity to effectively learn the material. This is where the prescriptions from the pacing chart can actually align with TARGETTS to promote adaptive motivation. By depicting the key ideas for units, the pacing chart offers insights into where teachers can address multiple mathematical goals by using fewer problems.

TARGETTS and Adaptive Motivation in the Mathematics Classroom

During professional development workshops on motivation, we use TARGETTS to highlight tangible steps that K–12 mathematics teachers can take in their daily lesson planning to promote adaptive motivation. Evaluations from teachers often state that the TARGETTS framework is the most important thing that they learn in the workshop: "[The most important thing I learned was] TARGETTS—I can use this as my basis for evaluating and making improvements to my day." Further, the following comments from teachers' evaluation forms explain the benefits of TARGETTS. By examining lessons they have already taught, teachers become familiar with the framework, learning to "[use] TARGETTS as a reflective tool." They also see where they are on track and where they can adjust to further motivate their students. Once teachers are familiar with TARGETTS, we encourage them to use the framework in their lesson planning, allowing teachers to "include all aspects of TARGETTS into a lesson and evaluate each aspect." Gilbert and Musu (2008) offer a more detailed discussion of specific practices associated with adaptive motivation for each TARGETTS dimension as well as classroom examples that middle and high school mathematics teachers generated for each TARGETTS dimension.

Using TARGETTS to Analyze a Lesson

During our professional development workshops, teachers describe and then analyze a recent mathematics lesson using TARGETTS to consider how instructional decisions might affect students' adaptive motivation. As you read the vignette from Lisa's eighth-grade algebra class, consider how the lesson addressed each TARGETTS dimension and what aspects of adaptive motivation the teaching approach may have diminished or enhanced.

> Lisa Johnson's class had reached the end of chapter 3 (inequalities). Lisa's pacing guide gave her one day for a chapter review. She had photocopies of the practice test that her department chair distributed. At the beginning of class, Lisa reminded students that the chapter 3 test would be the next day. She handed out the practice test, saying, "Work individually. Raise your hand if you have a specific question. Whoever finishes first will get to read the answers to the class."
>
> Lisa walked around the classroom to check that students were staying quiet and on task. Susan raised her hand and asked, "Ms. Johnson, will all these things be on the real test? I only want to know what we have to study to get a good grade."
>
> "Everything on the practice test could be on the test tomorrow," Lisa responded. Maria was the first student to finish. She gave her practice test to Lisa, and they reviewed it together against the answer key.
>
> "Well done, Maria. You got them all correct!" Lisa said. When there were ten minutes left in class, Lisa called out, "Time's up. I know that some of you haven't been doing your homework, so you're probably not done yet, but now Maria is going to read the answers aloud, so try to pay attention." As Maria read the answers from the answer key, students marked whether they got the answer correct or incorrect. Some students who weren't finished copied down what Maria read. Some seemed to ignore her, doodling on their practice tests. After Maria finished, Lisa asked, "Are there any questions?" Several hands went up just as the bell started ringing. "Come in and see me at lunch. You really should have asked me during work time. Now we're out of time." Lisa noticed that students seemed unhappy as they left. "I just hope I don't fail tomorrow," Lisa overheard Marcus say to Susan.

Although Lisa wanted to support students' focus on understanding the material (mastery goals) and enhance their self-confidence in and value for the content, Lisa realized that what she heard from her students showed that her instructional decisions had actually diminished their adaptive motivation. For example, Susan's question and Marcus's comment suggest that the lesson emphasized performance rather than developing understanding, as would be desired to promote adaptive motivation.

Table 19.1 summarizes Lisa's analysis of her lesson using the TARGETTS framework. By giving students a practice test to complete individually (task), she discouraged them from helping one another and asking questions if they did not understand a problem (grouping, teacher expectations). Similarly, by allowing individual questions only until the end of the lesson, she limited the number of potential mathematical misunderstandings and student questions that she could have addressed, as the many hands raised at the end of the class period (time) showed.

Lisa's story of a chapter review is familiar to many teachers. Understanding how a typical lesson such as this may diminish students' adaptive motivation is a first step to using TARGETTS to improve a lesson.

Using TARGETTS to Plan a Lesson

We now present a vignette of a lesson by another eighth-grade teacher integrating the aspects of TARGETTS in ways that positively affect students' adaptive motivation. This time, the teacher used the TARGETTS dimensions when planning her lesson. We can follow along with table 19.1 again to see how Tina's decisions promoted students' adaptive motivation.

> Tina's algebra class reached the end of chapter 6 (systems of equations and inequalities), and she planned a day of review before the students would have to take the exam. Tina began by considering her *task* and her *time* constraints. She looked at the chapter test and realized that her students had become so bogged down by different solution approaches (e.g., graphing, substitution, elimination) that they were forgetting that all these approaches were strategies that they could apply to the same system of equations as well. She looked at the textbook practice test, noting there were four "naked numbers" problems for each of four solution approaches and then four story problems. Instead of assigning all twenty problems, she picked the four story problems because they focused students on integrating their knowledge of the different solution approaches. Tina started class, saying, "We have a test tomorrow, and we will spend today reinforcing your understanding of the ways we've learned to solve systems of equations and inequalities. I want you to solve each system by using two approaches. Make sure that, at some point, you use each of the techniques we have learned in chapter 6. Even if you don't have time to completely solve all the problems, at least think about how you would set each one up."
>
> Next, Tina thought about *autonomy* and *grouping*, especially in the context of reviewing for a test. She told students, "Work on your own for fifteen minutes and then you will have a chance to compare and justify your thinking

Table 19.1

Examining lessons with TARGETTS: How did the lesson address each TARGETTS dimension? Did you diminish or enhance students' adaptive motivation?

TARGETTS Area	Focus	Lisa's Review	Diminish Adaptive Motivation	Tina's Review	Enhance Adaptive Motivation
Task	How learning tasks are structured: what the student is asked to do	Students complete a textbook worksheet, providing answers only	Focus is only on the answers, emphasizing performance	Teacher modifies the worksheet to focus on student difficulties and solution approaches	Focus on methods enhances self-confidence, mastery goals
Autonomy or responsibility	Student participation in learning/school decisions	Teacher decides who reads the answers from the teacher's edition	Students not given any choice in the activity; diminishes value	Students choose the solution approaches to use for each problem	Students can do their "favorite" way first; enhances value
Recognition	The nature and use of recognition and reward in the school setting	Only the student who finishes first is recognized, reads the answers	Quick finishers are rewarded, emphasizing performance	Students' reasoning and approaches are recognized	Focus is on *understanding* the different ways to approach the problem
Grouping	How and for what purposes students are grouped	No grouping; students work alone	Self-confidence diminishes as students struggle to understand material they feel like they should know	Students spend time working alone, in pairs, and sharing with the whole class	Students increase their self-confidence by comparing and justifying their methods
Evaluation	The nature and use of evaluation and assessment procedures	Students hear the correct answers but not how to get them or why they are meaningful	Only getting the right answer is important, emphasizing performance	Students evaluated for problem-solving efforts more than answers	Reasoning is the focus of the evaluation, emphasizing mastery goals

Table 19.1—*Continued*

TARGETTS Area	Focus	Lisa's Review	Diminish Adaptive Motivation	Tina's Review	Enhance Adaptive Motivation
Time	The allocation of time to different classroom tasks and activities	Students spend most of the class time working alone or listening to answers being read	Students given little reason to value the lesson; insufficient time to ask questions diminishes self-confidence	Fewer tasks designed to focus students on understanding the key concepts	Time to grapple with the material in an interactive way; enhances self-confidence
Teacher expectations	Beliefs and predictions about students' skills and abilities	Teacher interacts only with students who have questions about the worksheet	Feedback from the teacher is contingent on speed of completion, emphasizing performance	Teacher expects all to share their thinking with a peer and circulates to hear students' conversations	Teacher highlights solution approaches and student effort, increasing self-confidence and mastery goal focus
Social interaction	The nature of teacher–student and student–student relations	Very limited interaction in the classroom	Students' self-confidence diminished as they realize that Maria understands and they don't	Students take turns sharing answers and methods; interactions are math focused	Focus on everyone sharing and developing understanding increases self-confidence, mastery goal focus

with your neighbor." Turning to *teacher expectations* and *evaluation*, Tina thought about what to do while students worked in pairs and then how to have students share with the whole class. Because her focus was on the solution approaches, while students discussed with their partners, Tina walked around to each group and asked students to explain their choice of techniques for the different problems. She decided to look for two students whose conversation would highlight the connections between the approaches. Tina made a note in her lesson plan to *recognize* the students for their hard work to understand each other's approaches.

Amber and Michael were intently discussing one problem (Burger et al. 2008, p. 382): "The sum of the digits of a two-digit number is 11. When the digits are reversed, the new number is 63 more than the original number. What is the original number?" Michael said, "I don't understand the way that Amber got her answer. We both have the same answer, but we did the problem differently." Tina looked at their work and decided this was the conversation she was looking for.

She asked Amber and Michael to write up their solutions for the class. When they were ready, Tina said, "Amber and Michael will be sharing with us the solution approaches that sparked an interesting conversation because they didn't understand each other's ways at first. Did other pairs have that problem? Yes, that is what this review day is all about, helping each other to understand these different solution approaches for systems of equations since that is what chapter 6 is all about. Thank you, Amber and Michael, for sharing your insights with the whole class. As they share, I expect everyone else to pay attention and ask them questions *after* they are done speaking if you do not understand." Tina knew it was important to remind her eighth graders of her expectations for appropriate *social interaction*.

Amber went first. Using an interactive white board, she clicked on one color and used it to represent how she substituted $x + 7$ for y in the equation $x + y = 11$, leaving the original pieces of the equation a different color. "That was a very clear explanation. The different colors really helped us follow your thinking," Tina said to Amber when Amber had finished. Michael carefully wrote out his equations, showing how he used elimination. When he got to the step of $x + 7 = y$ and $x + y = 11$, he stopped and said, "Hey, wait a minute, I see another way to solve this one! I can just think of the second equation as $y = 11 - x$, and then I have $x + 7 = 11 - x$. That's another way to use substitution." Michael then proceeded to write out both his original method and his new method. "Your explanation helps us all to see how you can use both substitution and elimination to solve this problem," Tina commented to Michael. After Amber and Michael had both presented their solution approaches, she asked if anyone else in the class could show them another way. On the way out the door, Tina overheard Michael telling Amber, "I'm glad you presented the way you did the problem; it really helped me to understand another way to think about it."

Tina made small revisions to the supplied chapter review that still occurred within a class period (time), but she accomplished a great deal in promoting adaptive motivation. She encouraged students to focus more on learning and understanding the material by altering the *task*, reducing the number of problems to a set that specifically addressed earlier difficulties, and adding a section for students to explain their answers. She also enhanced students' value for the different approaches by setting the *teacher expectation* that they would use multiple ways and *recognizing* and *evaluating* students in the context of their approaches. Tina increased students' self-confidence through her decisions related to *autonomy*, *grouping*, and *social interaction*. She allowed students to decide which problem to use the different approaches for (autonomy) and permitted them to privately review their work with a partner (grouping) before having Amber and Michael carefully explain their solutions to the class (social interaction).

Reviewing the two vignettes in light of table 19.1, we see how the decisions that teachers make each day can diminish and enhance students' adaptive motivation. Although Lisa and Tina were both conducting a chapter review and had similar objectives, their decisions affected the outcomes for their students, in both motivation and likelihood for success on the upcoming test. In an environment of greater accountability with concerns about getting students through the content before high-stakes exams, the tendency is to think that Lisa's more traditional teacher-directed instruction will be more efficient than Tina's more reform-oriented, student-directed approach. However, the data in figures 19.1 and 19.2 illustrate that students who are more adaptively motivated in math are higher achieving in standardized test performance and in their ability to demonstrate adaptive reasoning.

Conclusion

In this article, we defined adaptive motivation and gave examples from recent studies that illustrate why it is important for mathematics achievement. We then shared TARGETTS and showed how this research-based framework is rooted in the literature on classroom practices associated with adaptive motivation. Finally, we presented vignettes from two eighth-grade mathematics classrooms that show how teachers can use TARGETTS to make small changes to their curriculum to enhance students' adaptive motivation. Regardless of students' initial attitudes or the constraints of the mathematics curriculum, TARGETTS offers a framework that mathematics teachers can use to support adaptive motivation.

(We are grateful to the National Science Foundation [grants 0335369 {Motivation Assessment Program}, 0314959 {TEAM-Math}, and 0227303 {TASEL-M}] for supporting our professional development efforts, and we thank Marty Maehr, Stuart Karabenick, Liz De Groot, Lisa Goldstein, and the editorial review panel for helpful comments and suggestions on earlier drafts of this paper.)

REFERENCES

Ames, Carole A. "Motivation: What Teachers Need to Know." *Teachers College Record* 91 (Spring 1990): 409–21.

Ames, Carole A., and Jennifer Archer. "Achievement Goals in the Classroom: Students' Learning Strategies and Motivation Process." *Journal of Educational Psychology* 80 (September 1988): 260–67.

Anderman, Eric M., and Chris A. Wolters. "Goals, Values, and Affect: Influences on Student Motivation." *Handbook of Educational Psychology.* 2nd ed. Edited by Patricia A. Alexander and Philip H. Winne, pp. 369–89. Mahwah, N.J.: Lawrence Erlbaum Associates, 2006.

Burger, Edward B., David J. Chard, Earlene J. Hall, Paul A. Kennedy, Steven J. Leinwand, Freddie L. Renfro, Tom W. Roby, Dale G. Seymour, and Bert K. Waits. *Algebra I.* Austin, Tex.: Holt, Rinehart, and Winston, 2008.

California State Board of Education. *Mathematics Content Standards for California Public Schools: Kindergarten through Grade Twelve.* Sacramento, Calif.: California State Board of Education, 1999.

Eccles, Jacquelynne, Terry F. Adler, Robert Futterman, Susan B. Goff, Caroline M. Kaczala, Judith L. Meece, and Carol M. Midgley. "Expectancies, Values, and Academic Behaviors." In *Perspectives on Achievement and Achievement Motivation,* edited by Janet T. Spence, pp. 75–146. San Francisco: Freeman, 1983.

Eccles, Jacquelynne S., and Allan Wigfield. "In the Mind of the Actor: The Structure of Adolescent Achievement Task Values and Expectancy-Related Beliefs." *Personality and Social Psychology Bulletin* 21 (March 1995): 215–25.

Epstein, Joyce L. "Effective Schools or Effective Students: Dealing with Diversity." *Policies for America's Public Schools: Teachers, Equity & Indicators.* Edited by Ron Haskins and Duncan Macrae, pp. 89–126. Norwood, N.J.: Ablex Publishing Corp., 1988.

Gilbert, Melissa C. "Applying Contemporary Views of Mathematical Proficiency to the Examination of the Motivation–Achievement Relationship." Paper presented at the annual meeting of the American Educational Research Association, New York City, 2008.

Gilbert, Melissa C., and Lauren E. Musu. "Using TARGETTS to Create Learning Environments That Support Mathematical Understanding and Adaptive Motivation." *Teaching Children Mathematics* 15 (October 2008): 138–43.

Jacobs, Jennifer K., James Hiebert, Karen B. Givvin, Hilary Hollingsworth, Helen Garnier, and Diana Wearne. "Does Eighth-Grade Mathematics Teaching in the United States Align with the NCTM 'Standards'? Results From the TIMSS 1995 and 1999 Video Studies." *Journal for Research in Mathematics Education* 37 (January 2006): 5–32.

Kilpatrick, Jeremy, Jane Swafford, and Bradford Findell, eds. *Adding It Up: Helping Children Learn Mathematics.* Washington, D.C.: National Academies Press, 2001.

Maehr, Martin L., and Eric M. Anderman. "Reinventing Schools for Early Adolescents: Emphasizing Task Goals." *Elementary School Journal* 93 (May 1993): 593–610.

Middleton, James A. "A Study of Intrinsic Motivation in the Mathematics Classroom: A Personal Constructs Approach." *Journal for Research in Mathematics Education* 26 (May 1995): 254–79.

Middleton, James A., and Photini A. Spanias. "Motivation for Achievement in Mathematics: Findings, Generalizations, and Criticisms of the Research." *Journal for Research in Mathematics Education* 30 (January 1999): 65–88.

Midgley, Carol, Avi Kaplan, and Michael Middleton, "Performance-Approach Goals: Good for What, for Whom, under What Circumstances, and at What Cost?" *Journal of Educational Psychology* 93 (March 2001): 77–86.

National Council of Teachers of Mathematics (NCTM). *Principles and Standards for School Mathematics.* Reston, Va.: NCTM, 2000.

Patrick, Helen, Lynley H. Anderman, and Allison M. Ryan. "Social Motivation and the Classroom Social Environment." In *Goals, Goal Structures, and Patterns of Adaptive Learning,* edited by Carol Midgley, pp. 85–108. Mahwah, N.J.: Lawrence Erlbaum Associates, 2002.

Patrick, Helen, Julianne C. Turner, Debra K. Meyer, and Carol Midgley. "How Teachers Establish Psychological Environments during the First Days of School: Associations with Avoidance in Mathematics." *Teachers College Record* 105 (October 2003): 1521–58.

Silver, Edward A., Vilma M. Mesa, Katherine A. Morris, Jon R. Star, and Babette M. Benken. "Teaching Mathematics for Understanding: An Analysis of Lessons Submitted by Teachers Seeking NBPTS Certification." *American Educational Research Journal* 46 (June 2009): 501–31.

Stipek, Deborah J., Julie M. Salmon, Karen B. Givvin, Elham Kazemi, Geoffrey Saxe, and Valanne L. MacGyvers. "The Value (and Convergence) of Practices Suggested by Motivation Research and Promoted by Mathematics Education Reformers." *Journal for Research in Mathematics Education* 29 (July 1998): 465–88.

Turner, Julianne C., Debra K. Meyer, Kathleen E. Cox, Candice Logan, Matthew DiCintio, and Cynthia T. Thomas. "Creating Contexts for Involvement in Mathematics." *Journal of Educational Psychology* 90 (December 1998): 730–45.

Turner, Julianne C., and Helen Patrick. "Motivational Influences on Student Participation in Classroom Learning Activities." *Teachers College Record* 106 (September 2004): 1759–85.

Wigfield, Allan, and Jacquelynne Eccles. "Expectancy–Value Theory of Achievement Motivation." *Contemporary Educational Psychology* 25 (January 2000): 68–81.

Chapter 20

Developing Prospective Teachers' Productive Disposition toward Fraction Operations

Juli K. Dixon
Janet B. Andreasen
George J. Roy
Debra A. Wheeldon
Jennifer M. Tobias

With the advent of standards-based mathematics instruction, children must develop a productive disposition toward learning mathematics. Similarly, teachers' own dispositions should allow them to believe "that they are capable of learning about mathematics, student mathematical thinking, and their own practice" (Kilpatrick, Swafford, and Findell 2001, p. 384). Many prospective teachers believe that learning mathematics is nothing more than following a set of rules in a prescribed fashion, which is contrary to the notion that mathematics makes sense. As a result, the mathematical experiences that prospective teachers encounter in teacher education programs are extremely influential in developing a *productive disposition*, or the "inclination to see mathematics as sensible, useful, and worthwhile, coupled with a belief in diligence and one's own efficacy" (Kilpatrick, Swafford, and Findell 2001, p. 5). This is particularly true for fraction operations. Prospective and in-service teachers encounter the same difficulties as children when learning fraction operations (Ball 1990; Tirosh 2000; Ma 1999; Borko et al. 1992). Another consideration hindering many prospective teachers' understanding is the stance that "if I, a college student, do not know something, then children would not be expected to know it, and if I do know something, I certainly do not need to learn it again" (Philipp et al. 2007, p. 439). In light of these issues, it becomes important to examine what experiences support the development of prospective teachers' productive disposition toward fraction operations by cultivating classroom experiences for learning about fractions.

Children's classroom experiences are based on several factors, including teachers' knowledge and beliefs related to teaching. Classroom instruction depends on how children engage with learning tasks. Pictorial representations, when used well, can give children the opportunity to link their informal knowledge and experience to mathematical abstractions (Kilpatrick, Swafford, and Findell 2001). Children who work with pictorial representations often develop a deeper understanding of fraction operations than children who have less exposure to these representations (Cramer and Henry 2002). "Students learn mathematics through the experiences that teachers provide. Thus, students' understanding of mathematics, their ability to use it to solve problems, and their confidence in, and disposition toward, mathematics are all shaped by the teaching they encounter in school" (National Council of Teachers of Mathematics [NCTM] 2000, pp. 16–17). Similarly, "how a teacher views mathematics and its learning affects that teacher's teaching practice, which ultimately affects not only what the students learn but how they view themselves as mathematics learners" (Kilpatrick, Swafford, and Findell 2001, p. 131). Though research has shown that children benefit from using pictorial representations, research has lacked in illustrating the effect on prospective teachers' productive disposition when they use such representations. In this paper, we propose that prospective teachers' use of pictorial representations along with an expectation that they explain and justify their mathematics solution processes will assist prospective teachers' development of productive dispositions, particularly sense making and diligence. We therefore trace how prospective teachers developed a productive disposition through classroom experiences focused on fraction concepts and operations.

Context

The prospective teachers that this article describes were part of a course focused on content for teaching elementary mathematics. Although mathematics for teaching was in the foreground of this course, modeling appropriate pedagogy held a prominent place in the background. The instructional tasks in the fraction operations unit were designed to support the development of prospective teachers' conceptual understanding of fraction operations by using pictorial representations where appropriate. The instructional experiences of these prospective teachers included opportunities to work with problems presented in context and to explain and justify their mathematical thinking. Furthermore, the instructor expected that the prospective teachers would use models, such as pictorial representations, to assist them in solving problems rather than relying on often poorly understood memorized procedures. The instructional choices were based on the belief that if prospective teachers were led to use pictorial representations to support their thinking and progress in computing with fractions, the representations would support their ability to make sense of the problems. The course's tasks followed Bruner's (1964) recommendations to begin with the enactive, or concrete, stage and to transition to the iconic and finally the symbolic. The instructor believed that using pictorial representations along with explanations and justifications to support them would thus foster the prospective teachers' productive dispositions toward fractions in particular. The work samples and whole-class dialogue that follow illustrate prospective teachers' products in the course.

The instructor initially introduced pictorial representations for fractions during the first session of an instructional unit focused on fraction concepts, so prospective teachers were familiar with representing various fractions with pictorial representations, including linear, area, and set models, before the fraction operations portion of the unit. Prospective teachers first encountered fraction operations through contextually based problems; an example from this instructional sequence was as follows: "Sue ate some pizza. Two-thirds of a pizza is left over. Jim ate $^{3}/_{4}$ of the leftover pizza. How much of a whole pizza did Jim eat?"

The course expected prospective teachers to use pictorial representations to make sense of mathematical situations such as this. Also, the instructor asked the prospective teachers to provide not only oral and written explanations but also justifications for their mathematical moves. Later, the unit transitioned from contextually based problems to computational problems that gave no context.

The following themes emerged from prospective teacher work samples, including explanations and justifications given in class discussions: (*a*) how pictorial representations allowed prospective teachers to see mathematics as sensible, useful, and worthwhile and (*b*) how pictorial representations assisted in prospective teachers' development of diligence and belief in their own efficacy as learners and future teachers of mathematics. The following sections will examine these two themes, beginning with describing initial dispositions and then by exploring how prospective teachers developed aspects of sense making and diligence.

Initial Dispositions

At the beginning of the semester, prospective teachers described their past experiences with and beliefs about mathematics. Initial beliefs that prospective teachers held about mathematics were often related to making mathematics sensible (or not), useful, and worthwhile. For example, three prospective teachers stated the following:

- "I think that it will be fun discussing the techniques and mnemonics that we were taught."
- "For me to understand math, it has to make sense. I struggled a little bit in my junior year of high school . . . because sometimes it just didn't

make any sense. But eventually I caught on to it. I also have to see the numbers to be able to work out the problem."

- "In my opinion, the only math most people use in life is what they learned in elementary school."

These examples are similar to those of Philipp et al. (2007), showing that prospective teachers initially believe that mathematics is about working with numbers or knowing procedures and demonstrate a perceived lack of need for understanding higher-level mathematics.

Developing Productive Dispositions

Prospective teachers went through three phases as they developed toward sense making. In the first phase, prospective teachers attached meaning to contextually based problems through pictorial representations. Establishing a norm that prospective teachers must be able to explain all mathematical procedures that they use in solutions pushed them into the second phase of development. The second phase included developing ways to explain and justify the mathematics through pictorial representations. The third phase consisted of developing new concepts and fostering productive disposition through making sense of the mathematics explored.

Phase One: Attaching Meaning

To aid prospective teachers' developing a more productive disposition toward mathematics, they were encouraged to make sense of contextually based problems by attaching meanings to the mathematics involved. For example, the "groups of objects" meaning from whole-number multiplication was attached to fraction multiplication. Specifically, prospective teachers built on meanings for operations with whole numbers as they began to explore fraction operations, in contrast to initial exposure to fraction operations that might focus on procedures for computing results. Prospective teachers in this class often interpreted problems by using the "groups of objects" meaning to make sense of fraction multiplication. For example, with the problem $^3/_4 \times {^2/_3}$, a "groups of objects" interpretation would involve finding $^3/_4$ of a group of $^2/_3$ of a whole. In figure 20.1, a prospective teacher used an area model to represent her solution process for this multiplication problem. Here the prospective teacher pictorially represented pattern blocks by using one hexagon as the whole, with two rhombuses representing $^2/_3$. Although not immediately, the prospective teacher then found that two rhombuses could be covered by four triangles, three of which would represent $^3/_4$ of the $^2/_3$, or $^1/_2$ of the hexagon. Through using a "groups of objects" model, the prospective teacher made sense of fraction multiplication.

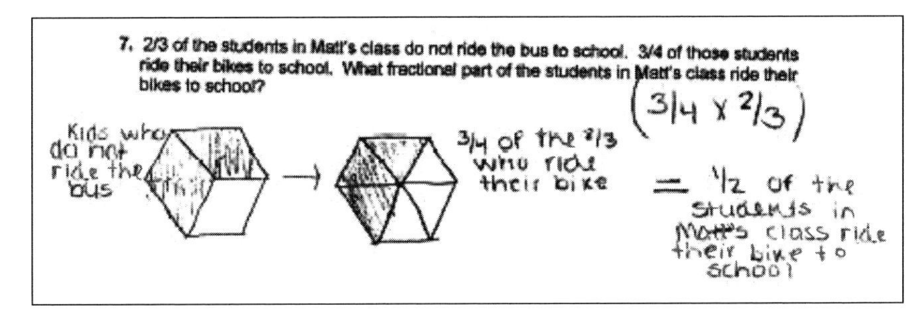

Fig. 20.1. Representation of $^3/_4 \times {}^2/_3$

As prospective teachers solved multiplication-of-fraction problems, the "groups of objects" meaning was pervasive; however, two distinct interpretations existed in solving division-of-fraction problems. The partitive (sharing) meaning for division was typically represented when the divisor was a whole number. For example, the problem $^3/_4 \div 3$ was most often interpreted as "If I share $^3/_4$ of something among three people, how much will each person get?" The measurement meaning for division was used when the divisor was a fraction. For example, the problem $3 \div {}^1/_4$, shown in figure 20.2, was interpreted as "If I have three of something, how many groups of $^1/_4$ of that something can be made from three?" Here the prospective teacher represented the problem linearly, beginning by representing three yards of ribbon and then dividing each yard into four pieces, each representing "$^1/_4$ yard long." The prospective teacher then counted the "12 pieces" that constituted the three yards of ribbon. Here the contextually based problem helped the prospective teacher make sense of dividing fractions.

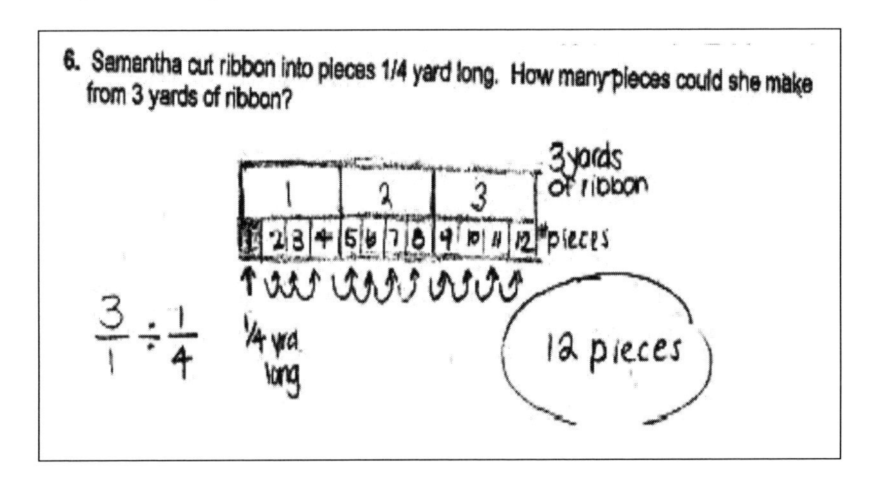

Fig. 20.2. Measurement meaning for division of fractions

This emphasis on attaching meaning to the division process was in line with Tirosh's (2000) findings that although prospective teachers could divide fractions procedurally, they could not explain their processes conceptually. The emphasis on the meaning of the division problems supported similar growth in these prospective teachers, which was evident in the representations that they used, fostering an increased productive disposition by making sense of multiplication and division of fractions in both contextually based problems and problems without context.

Phase Two: Developing Justification

Once the prospective teachers successfully attached meanings to a given problem situation, they developed ways to explain and justify their mathematical moves with pictorial representations. To explicate this process, we examined how one prospective teacher, Claudia, developed ways to explain and justify fraction multiplication with pictorial representations.

The instructor gave prospective teachers the problem $1^1/_3 \times {}^3/_4$, devoid of context, and asked them to develop a model to solve the problem. When prospective teachers were having difficulties representing the problem, the instructor encouraged them to make sense of the mathematics by thinking of the problem as a "groups of objects" situation. Even when the members of the class understood the problem as $1^1/_3$ groups of ${}^3/_4$, they still struggled to obtain and make sense of the correct answer. In the dialogue that follows, Claudia begins to develop ways of explaining and justifying multiplication in a "groups of objects" situation. Together the class first makes sense of the "groups of objects" model (lines 1–17), followed by Claudia's efforts to explain how the answer of 1 makes sense (lines 18–30).

1	*Instructor:*	What if we thought about this as groups of objects? $1^1/_3$
2		groups of ${}^3/_4$. How might we approach it that way?
3	*Claire:*	You draw ${}^3/_4$, $1^1/_3$ times.
4	*Instructor:*	Draw ${}^3/_4$, $1^1/_3$ times. Okay.
5	*Caroline:*	Wait.
6	*Jackie:*	To figure out $1^1/_3$ times, what do you mean?
7	*Instructor:*	Well, there's one time of ${}^3/_4$. Are we okay with that?
8	*Class:*	Yeah.
9	*Instructor:*	What's a third time of ${}^3/_4$?
10	*Olympia:*	Don't you have to break it into pieces? Split the slices.
11	*Instructor:*	Do I?
12	*Olympia:*	Yeah, because you have to get three, wait.
13	*Instructor:*	What is ${}^1/_3$ of this?

14	*Caroline:*	A fourth.
15	*Instructor:*	A fourth?
16	*Caroline:*	Yeah, out of the $^3/_4$.
17	*Instructor:*	She didn't split anything up.
18 19	*Claudia:*	My answer's 1. . . . No, I know because I don't know I'm thinking of it as if $^1/_3$ is $^1/_4$, so then . . .
20	*Instructor:*	Is $^1/_3$, $^1/_4$?
21	*Class:*	No.
22	*Caroline:*	What?
23	*Claudia:*	Forget it.
24	*Instructor:*	No, keep going.
25	*Olympia:*	I understand what you're saying.
26	*Instructor:*	What is she saying?
27 28	*Edith:*	Well, because you showed the $^3/_4$ out of 1, and so a third of the $^3/_4$ would be $^1/_4$.
29	*Instructor:*	Claudia, is that what you said?
30	*Claudia:*	Yeah.

Even though the problem had no context, the prospective teachers created a pictorial representation (fig. 20.3) to make sense of multiplying fractions. In figure 20.3 we see one group of $^3/_4$ represented by $^3/_4$ of a circle and then $^1/_3$ of a group of $^3/_4$ of the circle represented by $^1/_4$ of the circle. In essence, the prospective teachers applied the distributive property to $1^1/_3 \times {}^3/_4$ by finding one group of $^3/_4$ of the whole and then $^1/_3$ of a group of $^3/_4$ of the whole. They then combined the $^3/_4$ and $^1/_4$ to get one whole. In making sense of this pictorial representation, the prospective teachers, including Claudia, developed an understanding of multiplication by using a "groups of objects" model. Also, members of the class demonstrated the beginnings of diligence in solving mathematical problems, as lines 18–24 show. This emphasis on diligence persisted as prospective teachers engaged in the next task.

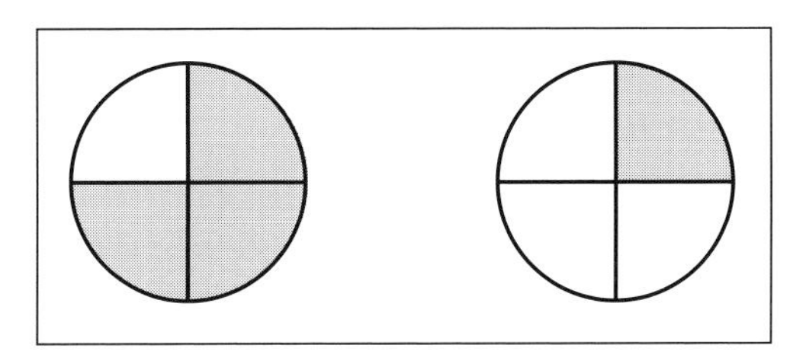

Fig. 20.3. Pictorial representation for $1^1/_3 \times {}^3/_4$

Phase Three: Fostering Diligence and Belief in One's Own Efficacy

To assist the prospective teachers in sense making *and* in developing diligence, the instructor posed another non–contextually-based problem of $1^1/_5 \times 1^2/_3$. During discussion of this problem, Claudia came to the board to explain how to find $1^1/_5$ groups of $1^2/_3$. Claudia continued to demonstrate sense making; however, and possibly more important, she also began to demonstrate diligence. Claudia introduced the concept of a "new whole," emphasizing that the whole changes throughout a multiplication problem.

1 2 3 4 5 6 7	*Claudia:*	So if we start off with two of these and divide them into thirds and then we find $1^2/_3$ . . . right? So then it would be this one, this one, this one, that's one. And then $^2/_3$ would be this much. So let's, like, draw because now $1^2/_3$ is our new whole because we're trying to find $1^1/_5$ of it (*fig. 20.4*). So if we just draw it altogether and make that our new whole. . . . That's right, right? Yeah. So then we look at this and we know (*fig. 20.5*).
8 9	*Instructor:*	How did she know that was right? What did she just check? You guys following her so far?
10 11	*Claudia:*	So I'm just combining these into here to make that our whole. Everybody follows?
12	*Class:*	Yeah.
13 14 15 16	*Claudia:*	Okay. So now this is our whole and now this is our whole and it's divided already into five. So then this is one and then one more would be $^1/_5$, right? Because this would be $^1/_5$, since this is our whole and then this is one piece of that.
17	*Suzy:*	I thought it had thirds; is that not a third?
18	*Claudia:*	Yeah this is a third of one thing. But then—
19	*Suzy:*	Each of those thirds right?
20	*Claudia:*	But then of this new whole, which is $1^2/_3$, this is $^1/_5$ of that.
21	*Suzy:*	All right.
22 23	*Claudia:*	. . . because it's divided into five. Does anybody not follow that?
24	*Alex:*	I'm not following it. Sorry.
25	*Suzy:*	Say it again. It took me a second, but that made sense.
26	*Claudia:*	This is what $1^2/_3$ are, right? Right here what I shaded.
27	*Suzy:*	Because we're finding $1^1/_5$ times.

28 29 30	*Claudia:*	And $1^2/_3$ is going to be our new whole, though, because we're trying to find $1^1/_5$ of that. So I just drew that as a new whole. Everybody okay up to there?
31	*Class:*	Yeah.
32	*Classmate:*	So you have that and then $^1/_5$?
33 34 35	*Claudia:*	Yeah, so then one more of these would be $1^1/_5$ because this is $^1/_5$, right, of this $1^2/_3$? And then one more would be that and it would be two (*fig. 20.6*).

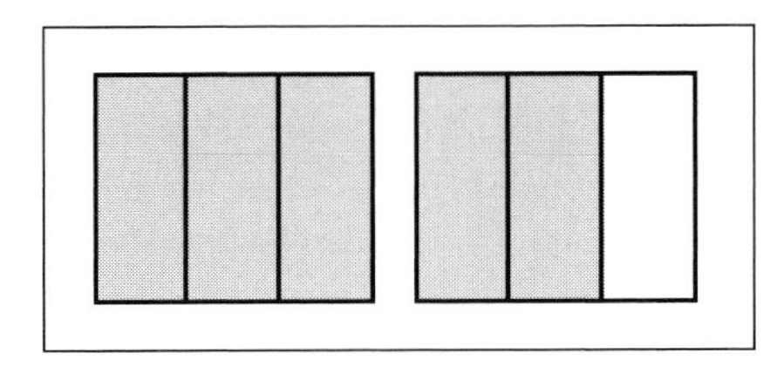

Fig. 20.4. Claudia's representation of $1^2/_3$

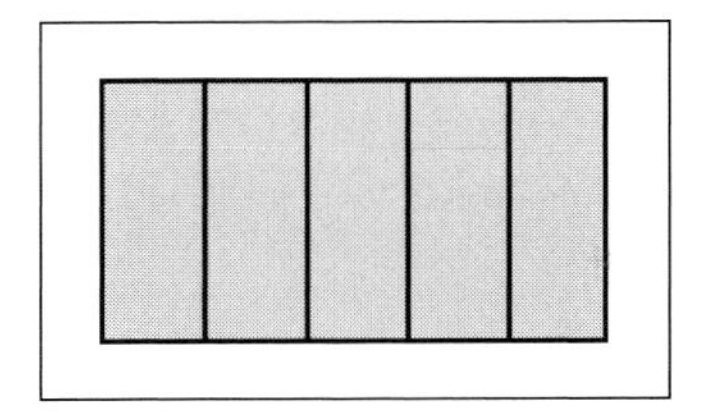

Fig. 20.5. Claudia makes a new whole.

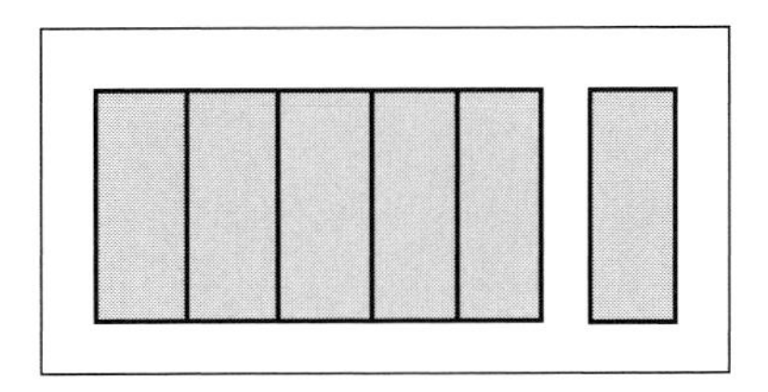

Fig. 20.6. Claudia's representation of $1^1/_5$ of $1^2/_3$

Claudia demonstrated diligence in supporting whole-class efforts to make sense of fraction multiplication. She did so by asking questions such as "Everybody follows?" (line 11), "Does anybody not follow that?" (lines 22–23), and "Everybody okay up to there?" (lines 29–30). Claudia's explanation was a result of her diligence in working to make sense of the mathematics. Without Claudia's diligence and belief in her own ability to learn mathematics, we believe that she would not have reached the level of sense making that the preceding exchange shows. Also, other members of the class, like Alex, began to demonstrate diligence in questioning Claudia's mathematical decisions to be sure that they were making sense of the mathematics correctly (lines 17–25).

Conclusion

The experiences afforded were designed on the premise that if prospective teachers faced contextually based problems—here related to multiplying and dividing fractions—and had to explain and justify their solution processes through using pictorial representations, they would make sense of the mathematics they encountered. Using pictorial representations supported prospective teachers' solution processes as they used the representations to explain and justify their mathematical moves. Through this process, prospective teachers developed productive dispositions toward fraction operations. Early in the unit on fraction operations, many teachers had difficulty performing computations as they struggled with attaching meaning to the situations presented. However, using pictorial representations helped them to make sense of the mathematics, correct their misconceptions, and develop more formal explanations of their mathematical reasoning. As their reflections at the end of the fraction unit show, their beliefs about mathematics also changed. The following comment indicates how their beliefs about mathematics and how to teach mathematics shifted from knowing procedures to the importance of knowing the meaning behind the numbers: "[As a teacher] I would come back to the word problem, because it's the word problem that explains what we're doing. I was never taught that. I hated word problems, and I think most people do because we weren't taught what we're really doing. I'd revert back to what the problem is saying and then show a visual. Once you get the meaning down, then you can talk about what it means to work with numbers."

This and other reflections from prospective teachers in the course demonstrated that using pictorial representations along with the expectation to explain and justify their mathematical moves facilitated developing productive disposition. This analysis has demonstrated how prospective teachers developed productive disposition—namely, making mathematics sensible, useful, and worthwhile and developing diligence and belief in their own efficacy as learners and future teachers of mathematics. Encouraging prospective teachers to use pictorial representations to help make sense of the mathematics when solving problems developed sense making and diligence.

REFERENCES

Ball, Deborah Loewenberg. "Prospective Elementary and Secondary Teachers' Understanding of Division." *Journal for Research in Mathematics Education* 21 (March 1990): 132–44.

Borko, Hilda, Margaret Eisenhart, Catherine A. Brown, Robert G. Underhill, Doug Jones, and Patricia C. Agard. "Learning to Teach Hard Mathematics: Do Novice Teachers and Their Instructors Give Up Too Easily?" *Journal for Research in Mathematics Education* 23 (May 1992): 194–222.

Bruner, Jerome. "The Course of Cognitive Growth." *American Psychologist* 19 (January 1964): 1–15.

Cramer, Kathleen, and Apryl Henry. "Using Manipulative Models to Build Number Sense for Addition and Fractions." In *Making Sense of Fractions, Ratios, and Proportions*, edited by Bonnie H. Litwiller and George Bright, pp. 41–48. Reston, Va.: National Council of Teachers of Mathematics, 2002.

Kilpatrick, Jeremy, Jane Swafford, and Bradford Findell, eds. *Adding It Up: Helping Children Learn Mathematics.* Washington, D.C.: National Academies Press, 2001.

Ma, Liping. *Knowing and Teaching Elementary Mathematics*. Mahwah, N.J.: Lawrence Erlbaum Associates, 1999.

National Council of Teachers of Mathematics (NCTM). *Principles and Standards for School Mathematics*. Reston, Va.: NCTM, 2000.

Philipp, Randolph A., Rebecca Ambrose, Lisa L. C. Lamb, Judith T. Sowder, Bonnie P. Schappelle, Larry Sowder, Eva Thanheiser, and Jennifer Chauvot. "Effects of Early Field Experiences on the Mathematical Content Knowledge and Beliefs of Prospective Elementary School Teachers: An Experimental Study." *Journal for Research in Mathematics Education* 38 (November 2007): 438–76.

Tirosh, Dina. "Enhancing Prospective Teachers' Knowledge of Children's Conceptions: The Case of Division of Fractions." *Journal for Research in Mathematics Education* 31 (January 2000): 5–25.

Chapter 21

Transforming Mathematics Teachers' Attitudes and Practices through Intensive Professional Development

W. Gary Martin
Marilyn E. Strutchens
Michael E. Woolley
Melissa C. Gilbert

The United States faces significant challenges in giving all its students a high-quality mathematics education. Although performance on international assessments such as the Trends in Mathematics and Science Study (National Center for Educational Statistics [NCES] 2008a) shows significant gains over the past decade or more, on the Programme for International Student Assessment (2007) U.S. fifteen-year-olds scored significantly below the international average in their ability to use mathematics to solve complex problems. Moreover, in the latest National Assessment of Educational Progress, less than one-third of eighth-grade students achieved the level of proficiency in mathematics, and less than one in seven achieved the "advanced" level (NCES 2008b).

These challenges are even more pronounced in Alabama, which ranked forty-ninth of the fifty states on the National Assessment of Educational Progress at grade 4 and tied for last among the fifty states at grade 8 (NCES 2008b). Only one in five Alabama eighth-grade students achieved at or above the level of proficiency, with one in fifty at the advanced level. As is true throughout much of the country, significant disparities in performance existed between white and African American students and between students of poverty and students not living in poverty. For example, only one in twenty-five African American eighth-grade students scored at the proficient level, compared with one in four white students. Thus, the mathematics performance across different groups of students needs improvement.

Auburn University, Tuskegee University, and fourteen school districts in east Alabama formed a partnership in November 2002 to face this significant challenge. This partnership received major funding from the National Science Foundation Math and Science Partnership program in 2003, along with several other internal and external grants. The mission for this partnership was "to enable all students to understand, utilize, communicate, and appreciate mathematics as a tool in everyday situations in order to become life-long learners and productive citizens by Transforming East Alabama Mathematics" (TEAM-Math 2003). This article focuses on the partnership's efforts to enhance student motivation and achievement by improving teachers' attitudes toward and use of reform practices (i.e., consistent with the recommendations of *Principles and Standards for School Mathematics* [National Council of Teachers of Mathematics {NCTM} 2000]). We first describe the partnership, then discuss specific aspects of the professional development, and finally share results from the first three years of the partnership.

TEAM-Math Partnership: Goals and Assumptions

A central goal of the TEAM-Math partnership is to ensure that *all* students, including African American and other historically underserved groups, receive high-quality mathematics education. Meeting this goal requires a comprehensive set of strategies addressing all aspects of the educational system. Thus, the partnership has been working to systemically change what is happening in mathematics education across the east Alabama region (Kim et al. 2001). The partnership design includes five primary components, including (1) curriculum alignment; (2) teacher leader development; (3) intensive professional development; (4) outreach to stakeholders, especially parents; and (5) improvement of teacher education. The TEAM-Math logic model in figure 21.1 illustrates these components.

The logic model suggests that involvement in professional development will lead to change in teacher attitudes toward and use of reform practices, which in turn will positively influence student motivation, ultimately leading to improved achievement in mathematics. Previous analyses of TEAM-Math project data (e.g., Woolley et al., forthcoming) showed that students who reported greater teacher use of reform practices, higher teacher expectations, and higher teacher standards showed more confidence and interest in mathematics and less math anxiety. Moreover, students with more desirable levels of motivation to learn mathematics performed better in mathematics, including standardized test scores and self-reported grades in mathematics. A direct relationship between teachers' use of reform practices and teacher expectations and students' performance in mathematics was also apparent (Woolley et al., forthcoming).

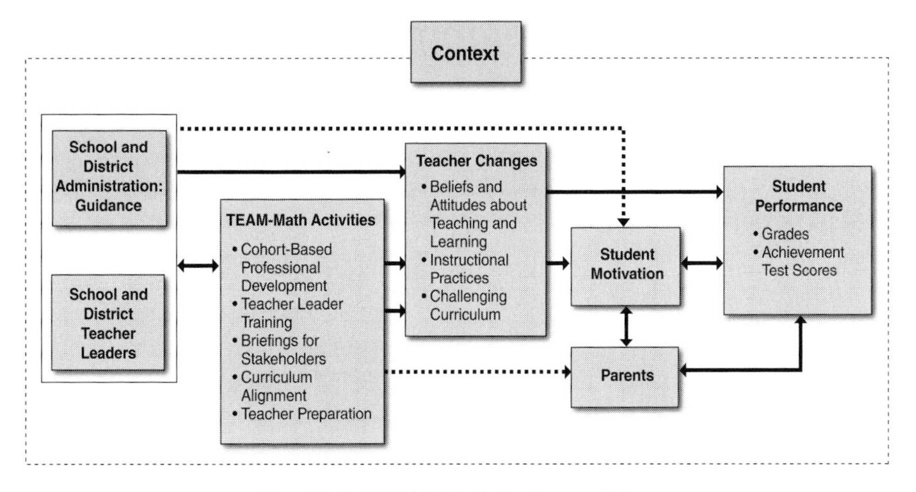

Fig. 21.1. TEAM-Math logic model

Here we focus on linkages in the model between teacher attitudes and instructional practices, which in turn affect student motivation and achievement.

Instructional Practices and Motivation

The teaching practices that TEAM-Math advocates are consistent with the findings of research focused on classroom strategies for enhancing students' motivation (e.g., Stipek et al. 1998; Stipek et al. 2001; Turner and Patrick 2004). However, an obstacle to implementing reform practices is teachers' own beliefs about mathematics teaching (e.g., Ross, McDougall, and Hogaboam-Gray 2002). TEAM-Math professional development activities were designed to affect teachers' beliefs about the nature of mathematics as a problem-solving activity and about what it means to learn mathematics, on the basis of national standards (NCTM 2000, 2006), state standards (Alabama State Department of Education 2003), and research on teaching and learning. Teachers were given opportunities to develop a variety of instructional strategies for students to explore curriculum content, a wide selection of sense-making activities or processes through which students can come to understand and "own" information and ideas, and many options through which students can demonstrate or exhibit what they have learned (Tomlinson 1995; Haberman 1992; Senk and Thompson 2003). Teachers were given an opportunity to enhance content knowledge through examining exemplary curriculum materials and solutions of "adult mathematics problems"—tasks that the teachers would find mathematically challenging. To address various expectations and levels of support for different groups of students, as stated in the Equity Principle (NCTM 2000), professional development activities challenged teachers to reconsider their beliefs about who can succeed in mathematics.

Professional Development

This section describes the professional development's structure, offers a sample plan for professional development, and discusses the professional development's hypothesized impact.

Structure

The structure of TEAM-Math's professional development was based on best practices (Loucks-Horsley et al. 2003; Borasi and Fonzi 2002). A cohort-based model was used, where teachers at a school entered the professional development as a group. For a school to join a cohort, at least 85 percent of the teachers who taught mathematics had to agree to participate. Qualitative analyses of participating schools have shown the importance of developing a supportive environment—including administrators and teacher leaders—in encouraging teacher participation in project activities (Strutchens et al. 2009). Also, the school teacher leader(s) and the administrators at each potential cohort school had to submit a plan demonstrating how their school would implement activities promoting the partnership.

Together, teachers from a school experienced a two-week and a one-week summer institute, quarterly follow-up meetings on Saturday mornings throughout the school year, other special workshops and events, and school-based activities focused on developing professional communities of practice (Wenger 1999). Figure 21.2 shows the components of a typical professional development day during a summer institute or a Saturday morning quarterly meeting.

TEAM-Math Challenge	TEAM-Math Challenges are "adult mathematical problems" that the participants will find challenging. They are solved individually or in groups, and then the solutions found by the individuals or groups are shared with the other participants, followed by reflection on pedagogical implications.
Pedagogical or Research Session	In pedagogical/research sessions, participants generally read and discuss articles related to research about mathematics education and/or pedagogical strategies.
Grade-Level Session	Participants look at activities, methods, and content issues particular to their grade or course level.
Reflections	Participants reflect over what they learned during the professional development workshop and relate new material to their teaching practices and previously learned materials.

Fig. 21.2. Typical TEAM-Math professional development components

Sample Professional Development Event

The following activity illustrates the TEAM-Math approach, taken from a quarterly follow-up meeting in 2005. All the teachers representing grades K–12 began the four-hour session by meeting in a full group setting, in which they were asked to work in pairs or trios to solve the "mangoes problem" while wearing their "student hats" (meaning that they were asked to focus on the mathematics in the problem).

> One night the king couldn't sleep, so he went down into the royal kitchen, where he found a bowl full of mangoes. Being hungry, he took $^1/_6$ of the mangoes. Later that same night, the queen was hungry and couldn't sleep. She, too, found the mangoes and took $^1/_5$ of what the king had left. Still later, the first prince awoke, went to the kitchen, and ate $^1/_4$ of the remaining mangoes. Even later, his brother, the second prince, ate $^1/_3$ of what was then left. Finally, the third prince ate $^1/_2$ of what was left, leaving only three mangoes for the servants. How many mangoes were originally in the bowl? (Stonewater 1994; NCTM 2008)

Teachers from different grade levels were asked to share their solutions for solving the mangoes problem to the full group. Figure 21.3 shows three such solutions. Discussion of the various approaches that teachers at different levels took reinforced the importance of encouraging multiple solution paths in the classroom. Then the teachers were asked to discuss the following questions, wearing their "teacher hats," in which they reflected on the pedagogical implications of the experience:

- What mathematical concepts would students need to know to solve the problem?
- What were the major mathematical concepts involved in the problem?
- What would students gain from solving this type of problem?
- What would students gain from presenting their solutions?
- What are some actions that teachers could take depending on how students respond to the problem?

On the basis of the teachers' experiences with the mangoes problem, a connection between research on differentiated instruction and how it relates to problem-based instruction as well as other major pedagogical emphases of TEAM-Math was made. Organized into grade-level breakout groups, teachers discussed how to implement differentiated instruction in their classroom through exploring a series of activities appropriate for their grade. The workshop ended with grade-level discussions of concepts and activities for the teachers to address in the next nine weeks. The approach taken illustrates the partnership's emphasis on balancing mathematics education research and theory with practical classroom concerns.

Working Backward: There are three mangoes at the end. That is half of what the third prince ate, so there were six mangoes before he ate. This is what is left after the second prince ate 1/3 of the remaining mangoes. But that is 2/3 of the mangoes before the second prince ate, so there were 9 mangoes before he ate. This is what is left after the first prince ate 1/4 of the remaining mangoes, which means this is 3/4 of the mangoes before he ate, so there were 12 mangoes before he ate. This is what is left after the queen ate 1/5 of the remaining mangoes, which means this is 4/5 of the mangoes before she ate, so there were 15 mangoes before she ate. Finally, 15 mangoes is what was left after the king ate 1/6 of the starting bowl of mangoes. So 5/6 of the starting bowl will be 15, which means that there were 18 mangoes to start.

Make a Picture:

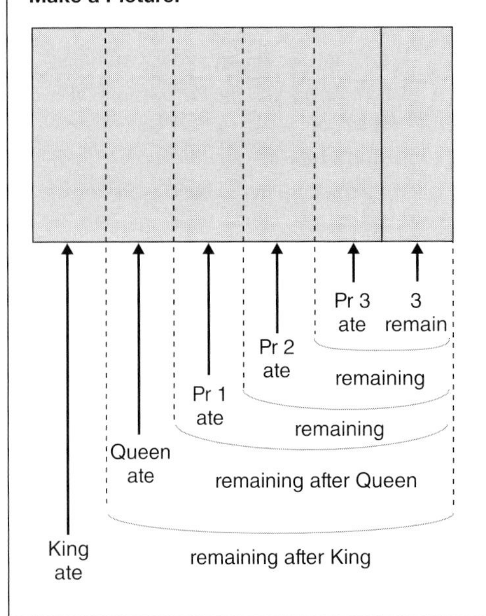

Guess and Check:
First try: Let's try 20. The king eats 1/6, so that doesn't come out evenly.

Second try: Let's pick a number divisible by 6, say, 24. The king eats 1/6, which is 4, leaving 20. The queen eats 1/5, which is 4, leaving 16. Prince 1 eats 1/4, which is 4, leaving 12. Prince 2 eats 1/3, which is 4, leaving 8. Prince 3 eats 1/2, which is 4, leaving 4. But we wanted 3.

Third try: We want a number a little smaller than 24, so let's try 18. Each person will eat 3, leaving us with 3 at the end. So that works.

Fig. 21.3. Typical responses to the mangoes problem

Hypothesized Impact

We have shown in previous analyses that teachers' use of reform practices in the classroom is related both to a more positive student motivation profile and to increased math achievement measured by standardized test scores (Woolley et al., forthcoming). In this paper, we test the other central aspect of the logic model described earlier and depicted in figure 21.1, that participation in TEAM-Math professional development activities will positively affect both teachers' attitudes toward and use of reform practices in the classroom. Moreover, we predict that participation in TEAM-Math will have a cumulative influence on the way teachers think about teaching and how they teach. That is, increased participation in the project, both during each year and accumulating across years, will lead to changes in both teachers' attitudes toward, and use in the classroom of, practices that the project advocates. In the following, we present an analysis of these hypotheses.

Teacher Sample and Analysis Plan

The sample for this analysis consisted of teachers from schools that participated in TEAM-Math for up to three years as of spring 2008. This sample included 655 elementary-level and 178 secondary school teachers from fifty-eight schools representing fourteen school districts in eastern Alabama. As far as years of involvement in the program, 353 teachers were from schools in their third year of involvement, 255 from schools in their second year, and 225 teachers were from schools in their first year. Of those who chose to report their demographic data, 749 teachers were female and 78 were male; 556 were white, 196 were African American, 25 were Asian American, 3 were Native American, 2 were Hispanic, 2 were biracial or multiracial, and 38 chose "other" race or ethnicity.

The analysis examines the relationship between participation in the preceding professional development activities and teacher attitudes toward—and ultimately use of—reform practices. Therefore, the analysis included three pivotal variables with respect to the goals of TEAM-Math. The first variable was the number of hours, based on project records of teacher participation in workshops, meetings, and other professional development activities that TEAM-Math offered. The second and third variables were self-report measures from a survey of teachers' attitudes toward and their reported use of reform practices in the classroom. This survey was given to teachers as a baseline in the spring before their school's participation in the first summer institute and was again administered each spring. We elaborate on each variable in the following.

Participation in TEAM-Math

Participation in TEAM-Math was measured by the number of hours each year (across the first three years of the project) that a teacher attended summer institutes, trainings, meetings, workshops, and other TEAM-Math activities. These hours were then recoded to reflect the level of *involvement* represented by the total number of accumulated hours across the years that each teacher participated. Therefore, hours accumulated during the first year of participation defined the level of participation for year one, the total hours of participation for years one and two were used to assign the involvement level at year two, and each teacher's total hours across all three years determined the year-three involvement level. Those levels of involvement were defined by various thresholds for participation hours in the project as follows: zero hours of participation was coded "none," one to thirteen hours was coded "minimal," fourteen to fifty-five hours was coded "limited," fifty-six to seventy hours was coded "involved," and seventy-one hours or more was coded "advanced."

Attitudes toward Reform Practices

Attitudes toward reform teaching practices were measured with five teacher self-report survey items completed in year three of the project. Teachers indicated their beliefs about reform teaching practices by their level of agreement with the following statements:

- Students should understand the meaning of a mathematical concept before they memorize the definitions and procedures associated with that concept.
- Teachers should allow students to figure out their own ways of solving mathematics problems.
- In a mathematics class, each student's solution process should be accepted and valued.
- Taking time to investigate why a solution to a math problem works is time well spent.
- Teachers should incorporate students' diverse ideas and personal experiences into mathematics instruction.

The survey questions had five response options for teachers to choose from, which ranged from "strongly agree" (4) to "strongly disagree" (0). These five items were summed to create a total score for each teacher, which could range from zero to twenty.

Use of Reform Practices

Use of reform teaching practices in the classroom was measured with four teacher self-report survey items in year three of the project. Teachers indicated their use of reform teaching practices by their level of agreement with how often students engaged in the following instructional activities in their classroom:

- Present to the class how they solved a problem.
- Take tests where they have to explain their answers.
- Participate in small-group discussions to make sense of mathematics.
- Think about why something in math class is true.

These survey questions also had five response options for teachers to choose from, which ranged from "all or almost all" (4) to "never" (0). These four items were summed to create a total score for each teacher, which could range from zero to sixteen.

Findings

We tested to see whether, as hypothesized, the level of participation in TEAM-Math was statistically related to teacher self-reports of attitudes toward and use of reform practices. Using SPSS version 15, we positively correlated the number of hours that a teacher participated in TEAM-Math professional development activities with his or her attitudes toward and use of reform practices. In fact, the correlations increased across the three years of teacher participation. For attitudes toward reform practices, in the first year the correlation with participation hours was $r = .15$ ($p < .001$); in year two, that rose to $r = .22$ ($p < .001$); and finally in year three, to $r = .29$ ($p < .001$). For teacher use of reform practices, the correlation with the first year of participation hours was $r = .16$ ($p < .001$); in year two, that rose to $r = .21$ ($p < .001$); and finally in year three, to $r = .28$ ($p < .001$). These significant statistical relationships are depicted in bar charts of the relationships between level of involvement in TEAM-Math and teachers' reported attitudes toward (fig. 21.4) and use of reform practices (fig. 21.5). In both charts, as the level of involvement by teachers increased from *None* through *Advanced*, teachers' attitudes toward and use of reform practices increased.

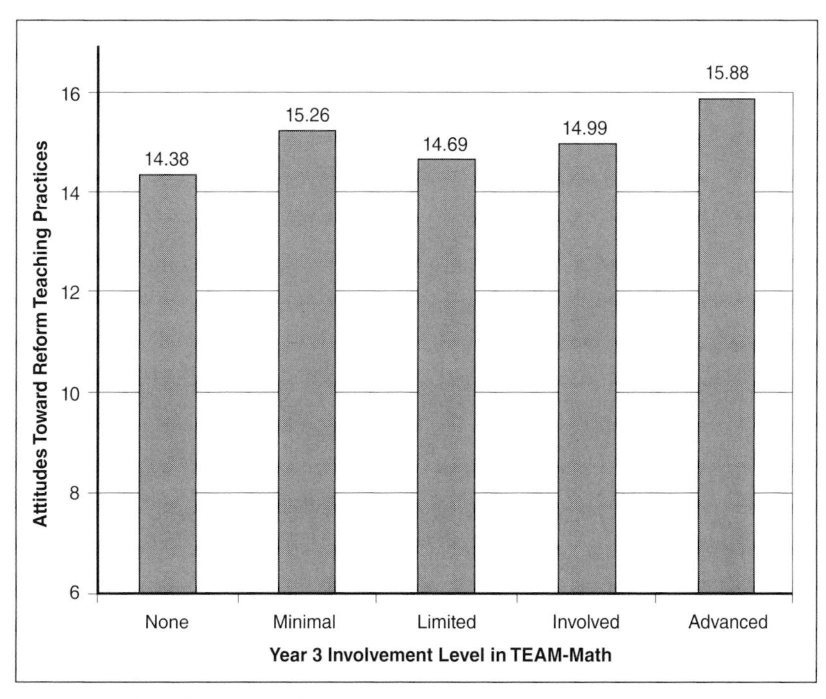

Note: Sample included 833 teachers across fifty-eight schools; involvement levels are determined by total number of hours of professional development over three years.

Fig. 21.4. Involvement in TEAM-Math and attitudes toward reform practices

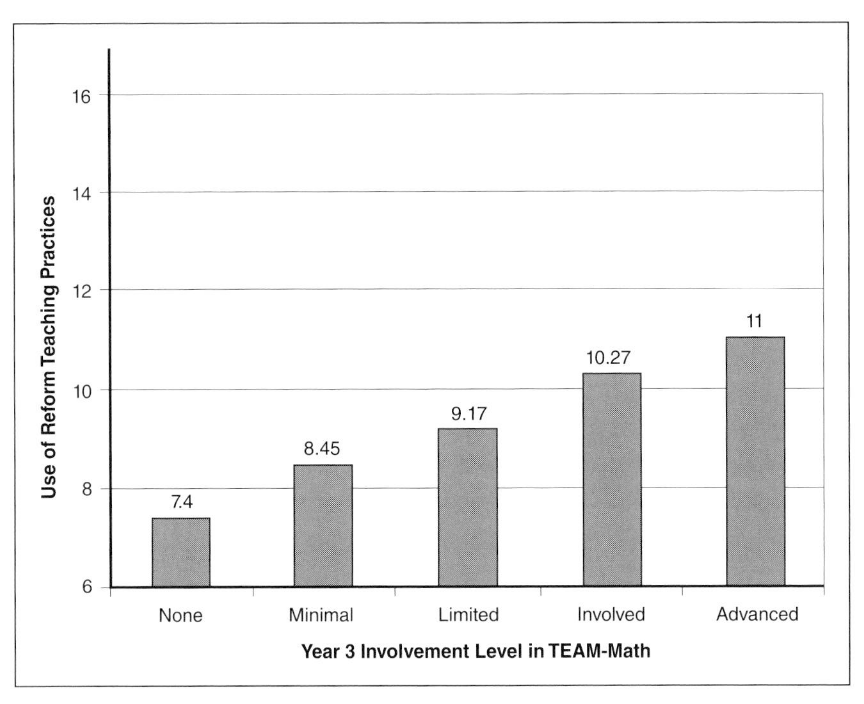

Note: Sample included 833 teachers across fifty-eight schools; involvement levels are determined by total number of hours of professional development over three years.

Fig. 21.5. Involvement in TEAM-Math and use of reform practices in the classroom

Conclusion and Implications

The analyses in the previous section suggest that teachers' participation in comprehensive professional development is positively related to their beliefs about reform practices and their use of those practices in their classrooms. This finding supports the TEAM-Math logic model in figure 21.1.

In these findings, we cannot easily disentangle causality between teachers' changes in attitudes and their changes in practices. That is, do changes in attitudes then cause changes in practices? Or do changes in attitudes occur as teachers change their practices? Although we cannot answer these questions, our analyses suggest that the two are interrelated. Thus, the TEAM-Math approach of simultaneously addressing attitudes and practices through more general explorations of the research and theory underlying reform mathematics—while also offering practical, grade-level guidance about how to change teaching practices—seems to be a reasonable approach to take and one that current and previous findings support.

Future analyses will attempt to link the two findings, as well as examine important factors that may affect teacher participation in professional development and indirectly influence changes in teacher attitudes and beliefs—including administrative support and parental support, as suggested in qualitative analyses (Strutchens et al. 2009). We plan further analyses in which we can triangulate teacher self-report with student reports.

Overall, our findings should encourage others interested in effecting improved mathematics teaching and learning—quality professional development that is both intensive and sustained leads to more teacher engagement, which leads to teacher change and ultimately to improved student motivation and achievement. Our experience suggests that teachers who are genuinely skeptical about reform mathematics teaching can be convinced of the utility of these methods through professional development that comprises three key features: (1) participating in mathematical challenges/adult problems with their "student hats" on, (2) engaging in investigative activities for their particular grade levels, and (3) reading and discussing articles that set forth the theoretical and research basis for changing practices. Professional development activities in which teachers experience their own mathematical growth in ways that are beneficial and eye opening help them to internalize the need for change in their instructional approach. After repeated personal experiences grounded in research and theoretical arguments, teachers report changes in their beliefs about and use of reform mathematics teaching practices. Even though teachers' beliefs about teaching and learning are hard to change, as are their teaching practices, we have seen that professional development experiences, such as the ones we described here, can cause teachers to change their teaching to more reform-based approaches. This change can in turn lead to their students' developing better attitudes toward mathematics and ultimately to improving achievement in mathematics.

(The National Science Foundation supported both Transforming East Alabama Mathematics [grant no. 0314959] and the Math Science Partnership—Motivation Assessment Project [grant no. 0335369].)

REFERENCES

Alabama State Department of Education. *Alabama Course of Study: Mathematics*. Montgomery, Ala.: Alabama State Department of Education, 2003.

Borasi, Raffaella, and Judith Fonzi. *Professional Development That Supports School Mathematics Reform* (Foundations Monograph No. 3). Arlington, Va.: National Science Foundation, 2002.

Haberman, Martin. "The Pedagogy of Poverty versus Good Teaching." *Education Digest* 58 (September 1992): 16–20.

Kim, Jason J., Linda M. Crasco, John Smithson, and Rolf K. Blank. *Survey Results of Urban School Classroom Practices in Mathematics and Science: 2000 Report.* Norwood, Mass.: Systemic Research, 2001.

Loucks-Horsley, S., Nancy Love, Katherine E. Stiles, Susan Mundry, and Peter W. Hewson. *Designing Professional Development for Teachers of Science and Mathematics.* 2nd ed. Thousand Oaks, Calif.: Corwin Press, 2003.

National Center for Education Statistics (NCES). *Highlights from TIMSS 2007: Mathematics and Science Achievement of U.S. Fourth- and Eighth-Grade Students in an International Context.* Washington, D.C.: NCES, 2008a. nces.ed.gov /pubsearch/pubsinfo.asp?pubid=2009001 (accessed June 28, 2010).

———. *The Nation's Report Card: State Mathematics 2008*. Washington, D.C.: NCES, 2008b. nces.ed.gov/nationsreportcard (accessed June 28, 2010).

National Council of Teachers of Mathematics (NCTM). *Principles and Standards for School Mathematics.* Reston, Va.: NCTM, 2000.

———. *Curriculum Focal Points for Prekindergarten through Grade 8 Mathematics: A Quest for Coherence.* Reston, Va.: NCTM, 2006.

———. "Classic Middle-Grades Problems for the Classroom." 2008. illuminations .nctm.org/LessonDetail.aspx?id=L264 (accessed June 28, 2010).

Programme for International Student Assessment. *PISA 2006: Science Competencies for Tomorrow's World.* Paris: Organisation for Economic Co-operation and Development, 2007. www.pisa.oecd.org/dataoecd/30/17/39703267.pdf (accessed June 28, 2010).

Ross, John A., Douglas McDougall, and Anne Hogaboam-Gray. "Research on Reform in Mathematics Education, 1993–2000." *Alberta Journal of Educational Research* 48 (Summer 2002): 122–38.

Senk, Sharon, and Denisse Thompson, eds. *Standards-Oriented School Mathematics Curricula: What Does Research Say about Student Outcomes?* Mahwah, N.J.: Lawrence Erlbaum Associates, 2003.

Stipek, Deborah J., Karen B. Givvin, Julie M. Salmon, and Valanne L. MacGyvers. "Teachers' Beliefs and Practices Related to Mathematics Instruction." *Teaching and Teacher Education* 17 (February 2001): 213–26.

Stipek, Deborah J., Julie M. Salmon, Karen B. Givvin, Elham Kazemi, Geoffrey Saxe, and Valanne L. MacGyvers. "The Value (and Convergence) of Practices Suggested by Motivation Research and Promoted by Mathematics Education Reformers." *Journal for Research in Mathematics Education* 29 (July 1998): 465–88.

Stonewater, Jerry K. "The 'Mangoes Problem.'" *Mathematics Teaching in the Middle School* 1 (November–December 1994): 204–10.

Strutchens, Marilyn E., Dan Henry, W. Gary Martin, and Lisa Ross. "Improving Mathematics Teaching and Learning through School-Based Support: Champions or Naysayers." Math and Science Partnership (MSP) Learning Network Conference, Washington, D.C., January 2009. hub.mspnet.org/index.cfm/lnc09_strutchens /page/index.

TEAM-Math. *2003 TEAM-Math Curriculum Guide*. Auburn, Ala.: Team-Math, 2003.

Tomlinson, Carol Ann. "Differentiating Instruction for Advanced Learners in the Mixed-Ability Middle School Classroom." *ERIC Digest* e536. Reston, Va.: ERIC Clearinghouse on Disabilities and Gifted Education, 1995. eric.ed.gov/PDFS/ED389141.pdf (accessed June 28, 2010).

Turner, Julianne C., and Helen Patrick. "Motivational Influences on Student Participation in Classroom Learning Activities." *Teachers College Record* 106 (September 2004): 1759–85.

Wenger, Etienne. *Communities of Practice: Learning, Meaning, and Identity*. Cambridge: Cambridge University Press, 1999.

Woolley, Michael E., Marilyn E. Strutchens, Melissa C. Gilbert, and W. Gary Martin. "Student Motivation and the Math Success of African American Middle School Students: Direct and Indirect Effects of Teacher Beliefs and Reform Practices." *Negro Educational Review* (forthcoming).